COMMUNITY MAPPING HANDBOOK

A Guide to Making Your Own Maps of Communities & Traditional Lands

By Alix Flavelle

Lone Pine Foundation
Distributed by Lone Pine Publishing
10145 – 81 Avenue
Edmonton, AB T6E 1W9
Canada

Website: http://www.lonepinepublishing.com

National Library of Canada Cataloguing in Publication Data

Flavelle, Alix, 1960–
    Community mapping handbook

    Includes bibliographical references.
    ISBN 1-55105-376-4

    1. Map drawing.   2. Cartography.   I. Title.
GA130.F52 2002   526   C2002-910837-3

Editorial Director: Nancy Foulds
Project Editor: Volker Bodegom
Editorial: Volker Bodegom, Genevieve Boyer
Research, Writing & Photography: Alix Flavelle
Field Trials & Vetting: Indigenous communities of Southeast Asia
Production Co-ordinator: Jennifer Fafard
Book Design: Volker Bodegom
Production: Volker Bodegom, Ian Dawe
Illustrations: Mary Brookes, Barb Turner, Lisa Kofod, Lorraine Gibson, Volker Bodegom
Cover Design: Elliot Engley

Every effort has been made to obtain permission to use all the information and examples included in the book. Any errors or omissions should be directed to the publisher for changes in future editions.

We acknowledge the financial support of the Government of Canada through the Book Publishing Industry Development Program (BPIDP) for our publishing activities.

TABLE OF CONTENTS

CHAPTER 4: PLANNING TO MAKE MAPS

CHAPTER 5: MAKING SKETCH MAPS

CHAPTER 6: PREPARING A BASE MAP

CHAPTER 7: USING TOPOGRAPHIC MAPS

CHAPTER 8: USING A TOPOGRAPHIC MAP WITH A COMPASS

CHAPTER 9: HOW TO DO A COMPASS SURVEY

CHAPTER 10: SURVEYING WITH THE GLOBAL POSITIONING SYSTEM (GPS)

CHAPTER 11: GATHERING LOCAL KNOWLEDGE ON MAPS

CHAPTER 12: PRODUCING THE FINAL MAP

CHAPTER 13: WHERE TO GO FROM HERE— USING THE MAPS

APPENDICES

SIDEBARS & TABLES

ACKNOWLEDGMENTS

This handbook is the outcome of the work of many people. Most of all, this handbook reflects the vision and efforts of the communities in Papua New Guinea, Thailand, Indonesia, Malaysia, and Nigeria with whom I have walked and mapped on their traditional lands. They allowed me the honour and privilege to work together with them to seek ways to put their holistic, multi-dimensional knowledge of the land onto flat paper, in a way that outsiders, such as myself, can understand.

Thank you to all the participants in the mapping workshops that I have facilitated over the years, for they are the ones who have, in their own communities, tried and tested the techniques presented in this handbook. And they have shown me the patience and persistence required to allow community process to unfold.

Thanks to Lisa Kofod, Mary Brookes and Barb Turner for the illustrative cartoons and to Lorraine Gibson and Volker Bodegom for computerized graphics contributions. Volker made the whole handbook consistent and readable with his meticulous editing and layout. Thanks to Shane Kennedy of Lone Pine Publishing for supporting this project from the idea stage.

Much of the writing was funded by the Environment and Development Support Program (EDSP) of the Canadian International Development Agency (CIDA) through the Endangered Peoples Project, and by the Biodiversity Support Program through the Indonesian Participatory Mapping Network. Editing, layout and printing were funded by the Lone Pine Foundation.

FOREWORD

This handbook is a guide for communities who want to make maps of their lands. It describes standard manual techniques that are used around the world for field mapping and drafting, and it provides a framework for how these techniques can be put together in a process called 'community-based mapping.' The material for this handbook has been compiled from experiences working with various indigenous communities, primarily in Southeast Asia, who in the last half decade have made great strides in mapping their traditional lands. Additional mapping methods have been inspired or adapted from mapping initiatives of First Nations in Canada. The emphasis of this handbook is on mapping indigenous territories, but the same methods can be used by any rural community interested in making maps that describe the lands that it lives on and the resources that it uses. The difference is in the content of the maps—what the community chooses to draw—rather than the basic techniques.

A community's immediate objective in making maps is often simple and symbolic: to affirm to the world its members' existence as a people by marking their place on the globe. In addition, communities around the world are using maps to

❖ *Document and preserve local/traditional knowledge about the land*
❖ *Plan and to manage community lands*
❖ *Raise community awareness about local land issues and to motivate communities to address them*
❖ *Increase local capacity to communicate and work with external agencies*
❖ *Assert aboriginal land rights in courts and negotiations*

Depending on its purpose, a community mapping project may involve just the most basic non-technical techniques—such as sketch mapping—or it may also involve more advanced techniques, such as compass surveys and GPS surveys.

The tools are pretty standard: paper and pencil are all that you need to begin mapping. And you can easily augment these tools with slightly more sophisticated aids—such as base maps, compasses, metre tapes, and GPS receivers—according to what is appropriate for the community's needs. What is unique in community mapping is who applies the techniques, the type of information being presented, and for what purposes.

In this handbook we acknowledge that the process of making maps is as important as the map product. By its nature, this mapping process serves a vital role in bringing community members together to walk on the land and to talk about the land. Therefore, it is not surprising that the most important ingredient for successful community-based map-making is not the equipment used. Rather, it is the full involvement of the members of the community.

Community members will come together to make a map
❖ *If they know that they can do it*
❖ *If they have something that is important to them to show on the map, and*
❖ *If they can envision how to use the completed map.*

Making maps of lands that a community has occupied for generations entails a long-term process, one that is different in each community. This handbook is intended to help communities to understand the basic principles of map-making and provide the information and tools necessary to tackle each step. Each community will then need to design for themselves a process that fits the community's unique culture, landscape, and purpose for making maps.

I hope that this handbook is useful in inspiring effective community-based mapping projects.

Alix Flavelle, August 2002

1 INTRODUCTION

1.1 ABOUT THIS HANDBOOK

This handbook is about map-making by rural communities that use and live on the land. It is about making maps of local knowledge. People who live on and depend on the land already have true maps in their minds. The hunter, the farmer, the medicine woman, the hobby naturalist and other people in the community already understand the pattern of rivers, the shape of the land, and the uses of the land. This handbook shows ways to draw these mental maps on paper in a way that everyone in the community, as well as outsiders, can understand.

Community mapping is sometimes called 'participatory mapping.' Capturing local knowledge on a map requires the participation and cooperation of the whole community—for instance, young people with good eyes and steady hands and the ability to learn new skills working with elders who can tell the stories of the ancestors and share their knowledge of the land.

Local people have pictures of their lands in their heads that can be translated into maps.

This handbook is oriented to applying standard mapping techniques to the unique endeavour of mapping specific customary lands. Many rural communities around the world have lived for generations on customary lands that are not officially titled.

Customary lands often
* ❖ *Are defined and managed according to local knowledge and traditional laws*
* ❖ *Have boundaries that follow natural features*
* ❖ *Have land uses, such as gathering forest products, that are seasonal or invisible*
* ❖ *Have landmarks that feature in stories that only the elders know*

What these communities have in common is a sense of connection to a homeland that affirms the inhabitants' history, culture and language, and that provides their livelihood.

Drawing local knowledge of customary lands on maps is different than making government maps or scientific maps. The history, culture, and land management practices of a people are written on the landscape of these lands. But an outsider does not know the history by looking at the land, and would have to be told the stories by community members. And, if the maps are to be an accurate reflection of the community lands, they must be made by community members.

Who Is This Handbook For?

This handbook is written for people who want to facilitate a community mapping project. The facilitator may be a member of the community that wants to make its own maps. Or the facilitator may be a staff person with a community development or indigenous organization that has been asked by the community to help facilitate a mapping project. This handbook is intended to assist the facilitator to help communities to draw accurate maps themselves. Ultimately, it is for communities who want to use maps to record and communicate information about the land on which they live.

How to Use This Handbook

This book begins by considering what maps are all about and the reasons that rural communities want to make maps. Then you will learn how to use simple tools to gather data about the land and draw it accurately on a map. A variety of standard techniques are presented. Since each community's needs are different, you will, in conjunction with the community, need to select and combine the methods most suited to your purpose.

This handbook contains a minimum amount of theory about maps, but enough to help you to understand the practical techniques that you need to know in order to make maps in a variety of situations. There cannot be one formula. This handbook can only give you guidelines to follow. It tells you what you need to know to make maps in your community, but it can't give you a recipe. Divided into small sections, this handbook is intended both as a guide to follow from beginning to end, and as a reference book in which you can easily find a topic or technique that you have a question about. You can look up definitions for a number of terms related to maps and map-making in the glossary at the back of this handbook.

Making maps involves first understanding some basic concepts about maps and then knowing how to employ a variety of techniques. Like learning anything, however, the most difficult part is knowing how to put it all together. It is best, therefore, to read this handbook through once so that you know how the sections fit together, and so that you have a picture in your mind of the whole process. Then practise. Get a base map and practice the survey techniques. Draw on it the information that you think is important.

Don't be intimidated by the technology and numbers. Think simple. Don't let the number of words or steps bother you. In mapping, the actual doing is faster than explaining the way to do it. Several words are often necessary to explain a single, simple action.

A mapping workshop in an Akha village in northern Thailand.

After practising mapping in the field yourself, you may be ready to train others and to be involved in designing maps. In order to train others, or to apply these techniques in many different communities, you must have a good understanding of basic mapping principles so that you can adapt to unique situations.

1.2 WHAT IS A MAP?

It has been said that maps are a universal language practiced by people everywhere. Even before the art of writing was developed, societies made maps, mainly for navigation. Their maps were made of materials from the forest: for instance, carved into wood or constructed from palm ribs.

The Pacific islanders made devices with frameworks of reed or palm leaves as navigation charts for sailing between islands. Aboriginal people in Alaska made

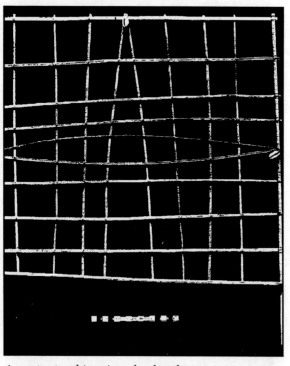

A navigational 'map' made of reeds.

wooden carvings to accurately depict parts of the Alaskan coastline.

Today, there are many types of maps that have been developed for different purposes. In general, a map is a picture or plan of an area, usually as viewed from above—as a bird in the sky might see it. Because this view differs from our usual ground-level view, we need to learn to look at maps from a different perspective than that to which we are accustomed.

On a map, as on a photograph, it is easy to see a wealth of information that would take pages of writing or hours of talking to describe. In another way, maps are not like photographs, which show everything on the land that can be seen

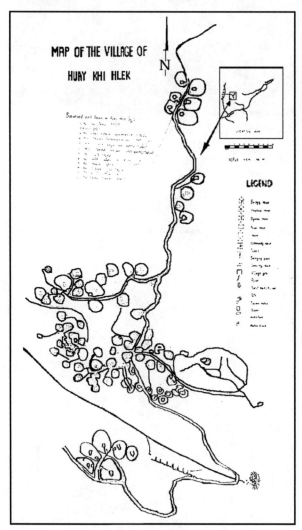

A village map made from a simple survey using a compass and metre tape.

from a certain place. Instead, maps use graphic symbols to show selected places and areas that might or might not be visible on a photograph, such as trails, hunting areas, gravesites, or historical places. From a map one can learn where any of these features is in relation to the others.

A map shows the position of features on the land and their spatial relationship to each other. For example, how far is the road from the river? How long and wide is the forest area or the garden? How far and in what direction is one mountain peak from another?

1.3   WHAT ARE MAPS FOR?

Maps have many uses. Here are five categories of uses for maps:

For recording and validating local or traditional knowledge

Maps provide an excellent way for a community to make a historical and cultural record of their way of living on the land. To start, it is important to record *how* and *where* the community uses its land and resources for food and shelter and living.

For example, the community members can show where their gardens are and where they collect food and medicines from the forest. It is also important to map their history on the land. They can show it using the place names, river names, spirit sites, legendary sites, and gravesites.

Maps are a visual way to record the locations of sacred sites and historical places.

For community organizing and awareness raising concerning land and environmental issues

Maps can also help to show how the land has changed over time, and to show how it might change in the future. What could happen to the land and the stories and the people when logging companies approach, when government schemes are offered, when a dam is built, or when people want to own their own farms within the village? How has the river changed since an industrial plantation was cleared above the village?

As the land (and life in the village) changes, maps can help everyone become aware of the changes and discuss the changes together. *How does the farming*

Maps are a way for villagers to discuss land-use issues.

activity of one family affect another? How does the water use in this community affect the water of the neighbouring community? All of these kinds of issues are more easily understood and discussed by everyone with the help of maps. Maps are a tool to initiate discussion within the community, to help people in the village to share their common experiences, to discuss the changes happening on the land, and to create a vision for the future of the land.

For planning and managing community land and resource use

Many indigenous communities have practised sustainable land-use management for generations. They never wrote a management plan and never needed to. Instead, the knowledge of how to manage the land was demonstrated or told to each generation by their parents and grandparents. Traditional institutions and laws coordinated activities on the land.

Disputes were mediated by a selected council or a council of elders according to ancient wisdom and traditional law. These traditional systems have always

Maps can help in the management of community lands.

been dynamic, but now they are changing rapidly under the pressures of industrial-scale resource extraction and national development schemes. Therefore, communities now use maps to help them to articulate their traditional 'management plan' to regional planners, or to make local land-use plans to improve their cash economy.

For increasing the ability of people in the community to communicate and work with external agencies

Maps are a visual tool for communicating with government officials, industry executives, or Non-Governmental Organization (NGO) advocates who come from afar and have influence to make decisions about local lands. By defining and recording local land use and occupancy in a standard way, it will help the community to talk to these 'outsiders' about the land. Because maps are a form of pictures, they are an effective way to communicate with people who do not speak the same language or who do not understand the local culture or what local people need. Government officials may consider the land to be empty and unproductive, but maps help to prove that this belief is wrong. Maps can show details about how the people use the land and the forest and the rivers. And instead of only the

Maps can help local people communicate their history and their vision to 'outsiders.'

few most educated people in the village speaking to outsiders, everyone in the village can understand and discuss a map.

For defining and negotiating customary land ownership

When an indigenous community surveys the location of its boundary and the locations of its historical sites, it defines the extent of its historical occupation and its priority on the land. In many countries of the world, the status of customary ownership is not legally clear. Just as maps are useful for discussions with outsiders about the management of the land, maps are also used for negotiating land ownership. In some countries, depending on the land laws, technical maps or historical maps are used in negotiations with government or submitted in court as legal evidence of ownership.

1.4   WHAT IS COMMUNITY MAPPING (PARTICIPATORY MAPPING)?

In community-based or participatory mapping, local community members make maps to describe the place in which they live. The people who live and work in a place have the most intimate knowledge of the place. Only they are able to make a detailed and accurate map of their history, land use, way of life, or vision for the future.

Participatory mapping differs from government mapping projects in many ways.

Community participation is key to making complete and accurate local maps.

Although the mapping techniques used in participatory mapping are standard to mapping in general, the difference is in *how* the techniques are applied and by *whom*. Because participation is so important, 16 villagers might work together to complete a compass survey that would normally require 2 people. Long into the night, groups of villagers may gather to draw and discuss the maps.

Another significant difference is that the villagers choose to draw maps with themes that are important to them. Maps of the community lands may already exist in government offices. But those maps show government land zonation, or road-engineering projects. Local communities tend to choose quite different themes, such as customary land boundaries, traditional farming practices, sacred areas, or fishing places.

So, what is special about the process of community-based mapping?

- ❖ *It involves everyone in the community in some way.*
- ❖ *The community defines its own issues and goals.*
- ❖ *The community directs the process.*
- ❖ *The mapping process, as well as the product, are designed to benefit the community.*
- ❖ *The majority of information on the maps comes from local knowledge.*
- ❖ *The community controls the use of the maps.*

Many foresters, agricultural extension officers, land-use planners, and development workers realize the necessity of doing community-based mapping for land-use planning and for resolving land-use conflicts. They are probably already familiar with sketch-mapping techniques as used in Participatory Rural Appraisal (PRA). The methods presented in this handbook are more rigorous in that villagers learn to make technical maps for themselves, and they are more participatory in the sense that the villagers themselves make the decision to produce the maps and request help to do so (if neces-

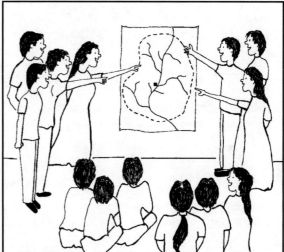

In participatory mapping, most everyone in the community gets involved in some way.

sary), decide what is drawn on the maps, and strategize about how the finished maps are to be used.

You've now reached the end of the introduction to this handbook. It should be clear to you what a map is and how important the local knowledge from the community is to making a map (or maps) of the community's choice. In the next chapter you will read about different types of maps and some of the basic principles of maps and mapping. The rest of this handbook is about how to plan and implement a mapping project with a community.

② PRINCIPLES OF MAPS

What you will learn in this section:
➢ *How to distinguish the various types of maps discussed throughout this handbook: sketch maps, scale maps, base maps, thematic maps and topographic maps*
➢ *The basic features seen on all to-scale maps*
➢ *How actual things (features) on the land are depicted with symbols on a map*
➢ *The importance of a legend to explain the symbols*
➢ *How the scale relates a measurement on the map to a distance on the ground*
➢ *How maps are oriented to north*
➢ *How a coordinate system allows you to locate yourself (or your map sheet) relative to the whole Earth*

2.1   TYPES OF MAPS

In this handbook we use the word **map** to mean 'a picture of the land.' A map can be a very simple, colourful picture drawn by hand and with nothing measured. Or a map can be full of lines, numbers and words, with everything on it measured. Maps can show any kind of information about the land, such as where in a region it rains the most or least, where in the forest the hunting is good, where the slopes are steepest

or which road to follow to get to a certain town. *Maps show where things are.* This section describes the various types of maps that you will read about in this handbook.

The **sketch map** is the simplest kind of map made in a community mapping process. Sketch maps are not drawn to scale. A sketch map is usually a freehand

The same scene shown as a photo, a panorama sketch, and a sketch map.

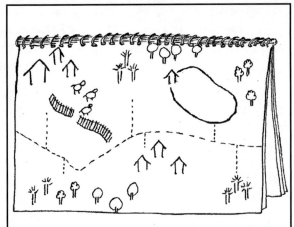

Making a temporary map on the ground allows many people to easily gather around to discuss and revise it.

drawing of the land on paper, using no tools for measuring, and drawn from the perspective of a bird looking down. However, a sketch of the land can be made using almost any materials and from any perspective. Some other types of sketches we call **ground maps**, **transects**, and **panorama sketches**. Read more about all these kinds of sketches in chapter 5.

A **scale map** is drawn to scale. It is a measured drawing or representation of the actual features on the ground. The distance between any two points on the map is in proportion to the distance between the same two points on the ground. For the distances to be consistent,

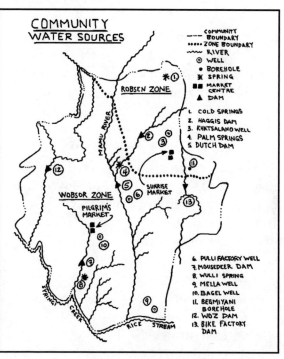

A sketch map can be drawn to show only a specified kind of information, such as water sources.

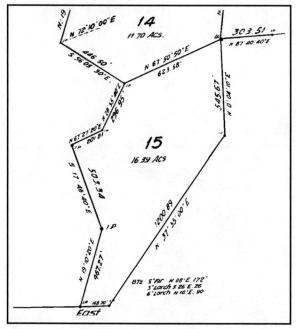

A land survey map made using a compass to measure direction and a metre tape to measure distance. The level of accuracy depends in part on the quality of these tools.

the direction must also be consistent. Any kind of subject can be drawn on a scale map. Read more about the meaning of scale in subsection 2.2.2. One example of a scale map is a **land survey map**.

A **base map** is a scale map that we use as a reference to draw the location of any other subject. For example, if you have a base map that shows the rivers of your territory, you can draw in your community's rice fields and rattan plantations in relation to the rivers. The rivers also provide a reference for you to draw the locations of the gravesites and sacred areas. The rivers are a standard reference for you to draw any aspect of your local knowledge. A river map to scale is just one kind of base map. Learn more about base maps in chapter 6.

A *topographic map* is a kind of map that is commonly used for a base map and for many other purposes. A topographic map is drawn to scale. It shows not only the rivers, but also features such as roads, and the shape of the land, indicating with contour lines exactly where the mountains, ridges and valleys are. Topographic maps are a standard kind of map made by government agencies. Learn more about topographic maps and how to use them in chapter 7.

To make *thematic maps* is the goal of a community mapping project. Thematic maps tell more of a story than a base map does. A thematic map can show any subject, for example:

❖ *community land use*
❖ *sacred areas*
❖ *local place names*
❖ *community use of forest plants*
❖ *farmed land in 1990*
❖ *land ownership*
❖ *vegetation types*
❖ *customary land boundaries*

We could combine a number of different subjects of the kinds above on one map, but usually doing so would make the map too cluttered. So we separate the subjects onto different maps. A series of thematic maps tells the story of the community. Thematic maps are normally drawn to scale by using base maps or by surveying or both. (However, a community sketch map could be a thematic map if it

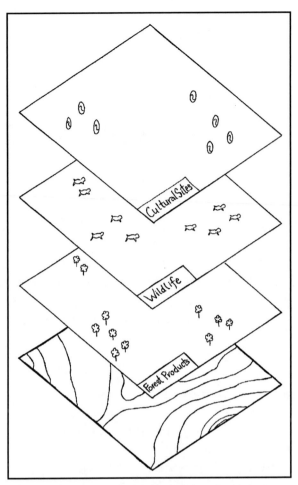

Three maps for the same area drawn with reference to the same topographic base map, but with each one emphasizing a different topic.

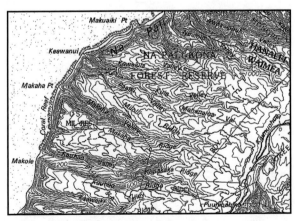

Local cultural information can be drawn on a topographic map.

shows information of the types listed above that is important to the community.) Chapter 4 describes how to design thematic maps for your community.

Through using this handbook, you will learn how to plan a mapping project (chapters 3 and 4), then to make simple sketch maps (chapter 5), and then to use the sketch maps to plan how to make accurate thematic maps using topographic base maps and land surveying (chapters 7–11).

But first, in the rest of this section, you will learn about the basic principles of maps and mapping. Scale maps have four basic elements:

❖ *Symbols to represent the different kinds of features on the map, plus a legend to explain what the symbols mean*
❖ *Scale to show the size and distance that the map represents on the ground*
❖ *Direction to show where north, south, east and west are on the map*
❖ *Coordinate System to describe where the mapped area is located on the Earth*

2.2   ELEMENTS OF MAPS

2.2.1   SYMBOLS

Maps are made using symbols. Symbols may be *point symbols*, *line symbols*, or *area symbols* (also called *polygons*). An example of a point symbol is a cross for a church. A dashed line that indicates a footpath is an example of a line symbol. An example of an area symbol is cross-hatched lines that depict a forested area.

Very common features—such as roads, rivers, mountain peaks and airports—have standardized symbols that are used by the majority of cartographers. Major roads are usually depicted with solid lines, whereas smaller roads, trails and boundaries are usually shown with different kinds of dashed lines. On a topographic map the elevation is shown by lines called *contour lines* (more information in subsection 7.2.2).

Special features related to a particular subject, especially on thematic maps, may be represented by symbols unique to a particular map or map set, because there are no standard symbols for these features. Because local knowledge about the land is unique to a particular place and culture, many aspects of it will require special symbols.

A *legend* is drawn in the margin of the map to explain the symbols that are used. Although many symbols are quite self-evident, for those that are not, the user can just refer to the legend to determine what they mean. Legends include written explanations of the symbols as well as information about the map itself, such as scale and north arrow (see below). Legends come in many shapes and sizes. Sometimes additional symbols are even printed on the back of the map sheet.

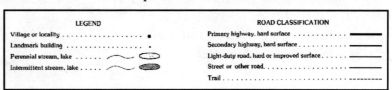

A legend describes in words the meaning of the symbols used on the map.

2.2.2   SCALE

For the purpose of consistency, most maps are made to scale.

When a map is to scale, it means that a certain distance measured anywhere on the map always represents the same distance on the ground. For example, if 1 cm on the map represents 1 km on the ground on one part of the map, then 1 cm represents 1 km anywhere else on the map.

Scale is commonly shown on the map written as a **fraction scale**, like this example:

1:25,000

which means

1 centimetre (cm) on the map = 25,000 cm on the ground, or

1 cm on the map = 250 metres (m) on the ground, or

4 cm on the map = 1 kilometre (km) on the ground.

Notice that this same ratio applies to other units of measurement as well, so

1 inch on the map = 25,000 inches on the ground.

The larger the number to the right of the '1:,' the less detail the map shows; we say that it is a **smaller scale** map. On the other hand, a smaller number means that the map gives more detail, and indicates a **larger scale**. That is, for a certain piece of land, a small-scale map (for example, 1:200,000) shows a small amount of detail—it shows everything small and close together and leaves off a lot of information. In contrast, a large-scale map (for example, 1:25,000) shows a large amount of detail—it shows everything larger and farther apart.

MAP TYPE	SCALE	LAND COVERED BY A MAP 40 CM × 40 CM
Large scale	1:5000 to 1:20,000	2 km × 2 km to 8 km × 8 km
Medium scale	1:25,000 to 1:150,000	10 km × 10 km to 60 km × 60 km
Small scale	1:200,000 and smaller	80 km × 80 km or larger

The scale can also be shown by a **graph scale**, also called a **bar scale**. This scale is like a small ruler that is printed in the margin of the map or in a featureless area of the map. As a matter of fact, a graphic scale can be used as a ruler that reads ground distance (in the units given) directly from the map. Graphic scales usually start at 0, but some extend to the left of 0. Be careful when you are using a graph scale that you start your measurement at 0 (unless you intend to use the extension, which is more finely subdivided than the rest of the scale so that you can use it to measure a distance more accurately than simply estimating).

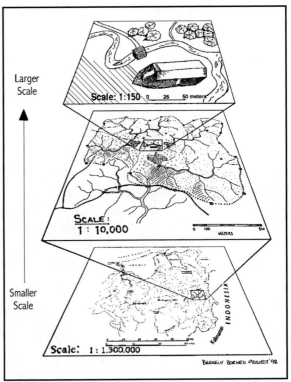

Maps of the same area shown at different scales.
(Used with permission of the Berkely Borneo Project.)

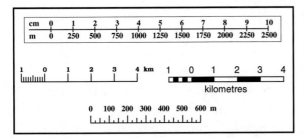

Graph (bar) scales can have a variety of formats.

Graph scales are more reliable than fractions. If someone photocopies a map and enlarges or reduces its size in the process, then the graph scale changes at the same rate and remains accurate. However, if the map only has a fraction scale, written as 1:50,000 for example, then you have to calculate the new scale on the photocopied map based on a known distance between two points on the ground, or based on the original map (or another accurate map).

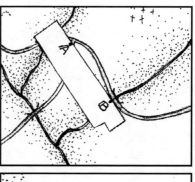

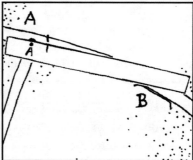

An easy way to use a graph scale is with a strip of paper, rather than with a ruler, because the marks on a ruler often do not correspond with the marks on the scale, which can be confusing. So, suppose that you want to know the distance between town A and town B. Take a blank strip of paper and lay it on the map so that one edge touches the points marking town A and town B. Simply make a pencil mark on the strip beside each town, and then lay the paper beside the graph scale and read off the number of kilometres between the two marks. (If the scale bar is too short, you will need to mark off a certain number of lengths of it on the paper and add them up.)

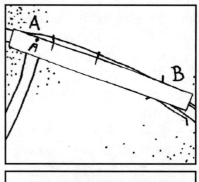

This method works well for curved lines also—for example, to measure the length of a river. Put a tick on the paper at the starting point on the map. Find the first bend in the river and make another tick, then pivot the paper until it lines up with next leg of the river and make a third tick at the second bend. Pivot the paper again and put a fourth tick at the third bend. Continue in that way until you have reached the end point of the distance to be measured. When you lay the strip next to the graph scale, the length of the river is the distance from the first mark to the last.

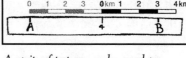

A strip of paper can be used to measure distance on a map, along straight lines and along curves. In the example at the bottom, the distance between A and B on the strip of paper as measured with the scale is 4 km + 3 km = 7 km.

If you know the fraction scale of the map, you can measure between any two points on the map and calculate the distance on the ground.

---

ground distance (cm) = map distance (cm) × scale
(where 'scale' is the number to the right of the '1:')

Or, again if you know the fraction scale of the map, you can measure the distance between two points on the ground and calculate the distance on the map.

---

ground distance (cm) = map distance (cm) × scale
(where 'scale' is the number to the right of the '1:')
conversely,
map distance (cm) = ground distance (cm) / scale

2.2.3 DIRECTION ON THE MAP

Maps are made to show directions as well as distances. Long ago, geographers developed a standard so that all maps show direction the same way and are provided with a **north arrow**. Moreover, maps are by convention drawn with north at the top of the map sheet (except some smaller ones that are drawn differently for reasons of design, such as in an advertising brochure). To understand the concept of drawing maps with north at the top, you have to remember that a map is a picture of a real place on Earth. The Earth rotates in space on an axis, with the north pole at one end and the south pole at the other. These poles are real points on the globe and give us a convenient reference for our maps. The whole reference system of directions and locations is based on these two points.

How do we know where north is? In the northern hemisphere it is marked by the north star at night. In the daytime, if it is not too cloudy, we can find the north and south poles by the angle of the sun. But what about the directions in between? Although you could use a simple protractor to measure them, many surveyors don't bother with the sun or stars at all. A simple and precise tool for measuring angles of direction is the **compass**, which works by magnetism, as will be explained later.

The Earth is surrounded by a magnetic field. The Earth's magnetic field is not exactly on the same axis as the north and south poles: *magnetic north* is near (but not the same as) the north pole and magnetic south is near (but not the same as) the south pole. The difference (or angle) between true north and magnetic north is called the **magnetic declination.** This declination varies from place to place, and can be either *west* or *east* of the north pole. For instance, in much of southeast Asia (Malaysia, Thailand, and western parts of Indonesia) the declination is quite small (less than 2°) and can usually be ignored when making all but the most precise maps. By contrast, the declination across North America varies from about 25° E to about 30° W, with even larger values possible north of 55° N.

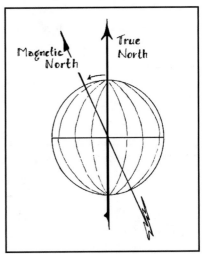

True north indicates the direction of the north pole. The magnetic declination is the deviation of the magnetic field from true north at a particular place.

If we can't see these magnetic lines, how do we know exactly where north is? Or where 30° from north is? We can stand anywhere on Earth and measure the angle from us to the magnetic north pole with a *compass*. A traditional (non-electronic) compass has a magnetic pointer that aligns itself with Earth's magnetic field.

We don't need a magnetic tool to measure the direction from one place to another on an existing base map, because the north lines are already assigned. We can use a simple protractor to determine the direction of a line between any two points simply by measuring the angle of the line in relation to the north lines. (See subsection 9.5.1, where plotting a traverse is described.)

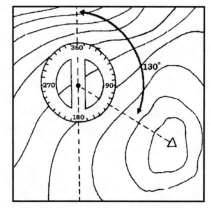

Direction on a map is measured, by agreement, as an angle from the north line.

2.2.4   COORDINATE SYSTEMS

If you don't know which area on Earth a map in your hand represents, it is just a piece of paper with some lines drawn on it that represent rivers, mountains, roads or political borders. But, unless you recognize familiar place names, how can you tell which area it represents? The use of a *coordinate system* provides a solution.

Professional map-makers have evolved the concept of imaginary *meridian lines* that run north–south (vertically) from pole to pole at standard intervals around the planet and *parallels* that run east–west (horizontally). These lines are numbered systematically (and the intervals can be subdivided in a predetermined manner if necessary). Any point on the surface of the Earth can be defined by a *coordinate* that indicates which vertical line and which horizontal line cross there. If you know the numbers assigned to at least one vertical line and one horizontal line that cross your map, then you know what little part of the Earth your map is a picture of.

There are several different kinds of coordinate systems. The oldest and most common is the *geographical coordinate*

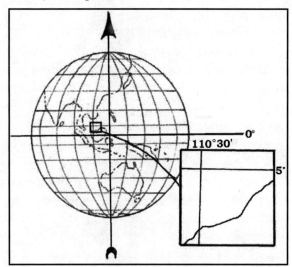

If your map has a standard coordinate system on it— commonly latitude-longitude or UTM—then the area covered by the map and any point on the map, can be precisely located on the globe.

system, which uses latitude and longitude. This system can be used for all basic navigation and surveying. Latitude (parallel east–west lines) and longitude (north–south lines connecting the poles) are numbered in degrees, minutes and seconds. Degrees of latitude are measured 0° to 90° N and S from the *equator*, which is the line that connects all points on the surface of the Earth that are equally distant from both poles. Degrees of longitude are measured 0° to 180° E and W of the *prime meridian*, which arbitrarily goes through Greenwich, England. Further subdivisions—minutes and seconds—are numbered 0 to 60 in each category.

Every point on the planet can be identified by its coordinates—where latitude and longitude lines cross. An example of a point in Kalimantan might be longitude 117° 45'15" E (117 degrees, 45 minutes, 15 seconds) and latitude 02°35'10" N (2 degrees, 35 minutes, 10 seconds). Longitude lines are aligned with true north, which is, in many places, in a slightly different direction than magnetic north. Look at the declination arrow in the margin of the map to determine the difference.

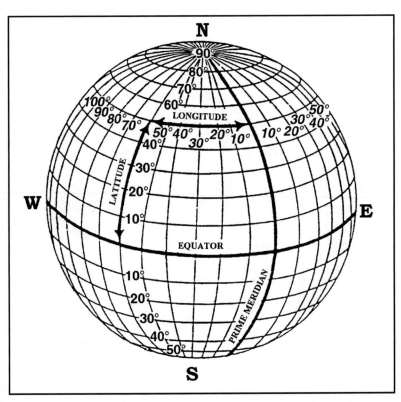

The geographic coordinate system with latitude and longitude meridians.

Another common kind of coordinate system is called a **rectangular coordinate system**. This kind of coordinate system is particularly good for navigating over long distances by calculating positions on a flat map of the Earth's curved surface. The **Universal Transverse Mercator** (**UTM**) grid system is one such scheme, and, in conjunction with the **Universal Polar Stereographic (UPS)** grid system for polar regions, it is widely used for topographic maps, air (aerial) photographs,

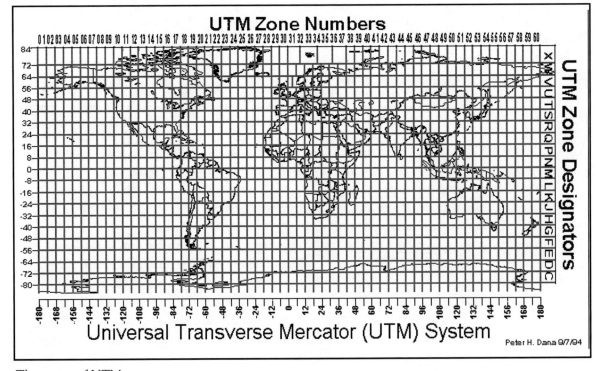

The system of UTM zones.

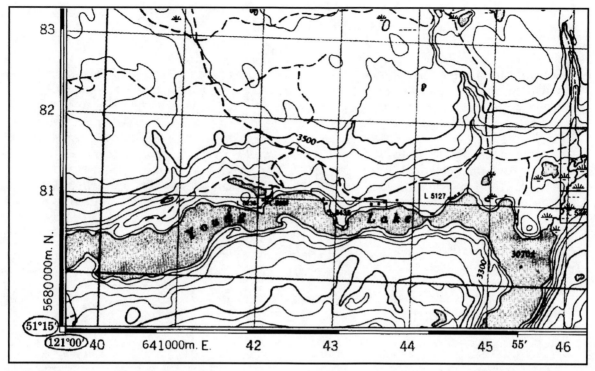

This map uses two coordinate systems, as is evident by the different numbering systems: longitude/latitude (circled) and UTM (at the grid lines).

and satellite images. In the UTM grid system, the area of the Earth between 84°N and 80°S latitude is divided into north-south columns, each of which is 6° of longitude wide. They are numbered from 1 to 60 eastward, beginning at the 180° meridian. These columns are subdivided from north to south into *zones*. These zones are lettered south to north with the letters C to X ('I' and 'O' are omitted), starting with 'C' at 80° south. Each measures 8° of latitude high (except zone X, which is 12°). Each 6° by 8° section of the grid (for example, 28M) covers a square that measures 100 km by 100 km.

On a UTM map you will see UTM grid lines running north–south and east–west. These north–south lines represent a third designation of north (in addition to true north and magnetic north), known as **grid north**. Grid north does not change relative to the features on the ground over time the way magnetic north does, so, for a particular place, the difference between grid north and true north is always the same.

A UTM map has numbers like $^{22}10^{000}$ in the margin (see appendix E). This grid is easy to use because it is directly related to metres and kilometres on the ground, according to the scale on the map. Easting coordinates are measured relative to the central meridian of the zone, which is assigned a value of 500,000 m (500^{000} or 50^{0000})

Sometimes you will see both UTM and longitude/latitude on the same map. Although having both may be confusing at first, note that the grids always have a different system of numbering and that the lines may be drawn in different colours or thickness so that they can be differentiated.

You are now familiar with the basic concepts of scale maps and are thus prepared to make scale maps. If you have never used these concepts in practice, they may not yet be very clear for you. Don't worry. As you try some of the techniques for making maps you will use these concepts and they will become more clear. You can always refer back to this section and read parts again.

GENERALIZATION

Generalization is the selection of features to show on a map and the simplification of how they are depicted so as to result in a clear and meaningful map. The kinds and degree of generalization are related to the **scale** and the **purpose** of the map. A map is a small drawing of a big area on the ground, and the scale of the map determines how much detail it is possible to draw. For example, on many maps a house must be drawn so small that it can be shown only with a very simple symbol, such as a dot or a small square. And, on a city map, it is impossible to draw millions of houses, so an area symbol is used to represent built-up areas. A decision to represent a house with a dot, rather than showing its true shape—or not to draw individual houses at all but to use a symbol for a settlement as a whole—is an example of the process of generalization.

The several fundamental processes that are used in generalization are applied differently to line, area, and point symbols.

Understanding the concept of generalization is especially important in map design. The decisions in map design are always subjective. There are three guidelines in the process of generalization:

☞ Generalization decisions must be made according to the purpose of the map

☞ Generalization decisions must be appropriate to the character of the area.

☞ The level of generalization must be consistent throughout a map series.

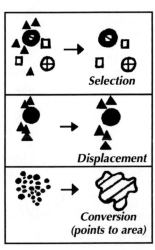

POINTS

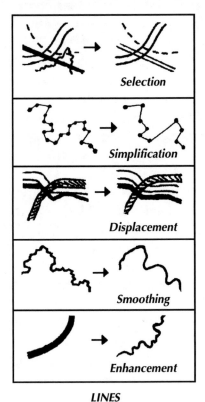

LINES

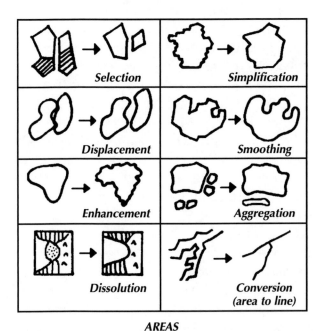

AREAS

MAP PROJECTION

Map projections are ways of portraying the surface of the Earth, or a portion of it, on a flat surface. When we try to represent a sphere on a flat map there is distortion. (Just try to flatten the skin of an orange!) Every map projection distorts four properties: shape, distance, direction, and area.

Projections are mathematically calculated. There are many types of projections and each one deals with distortion in a different way. For example, some projections minimize distortions in area and others minimize distortions of direction. The most important thing is that the coordinate system is correct. In other words, every point on the Earth has a relationship to parallel and meridian lines (or latitude and longitude lines), and the projection method must preserve these relationships.

In the projection process, the sphere is first converted to one of three flattenable surfaces: plane, cone, or cylinder. The choice of surface and the orientation of the sphere result in different projections.

In basic terms, the projection process works like this: First, the sphere representing the Earth is converted to one of three flattenable surfaces; plane, cylinder, or cone. Then, for each type of surface, the sphere can be oriented in a variety of ways—the poles may be aligned diagonally, horizontally, or vertically on the surface. Each different orientation of the projection surface produces a different pattern of map distortion, because the parallels and meridians line up differently.

Topographic maps and navigational maps in most parts of the world are made on the Universal Transverse Mercator (UTM) projection, because it maintains direction or angles. The Mercator projection is characterized by parallels and meridians that are straight lines and that are at right angles. This projection is based on a cylindrical surface with the equator as the axis. The Mercator projection uses the prime meridian as the axis of the map. Topographic maps of Indonesia, for example, are made on the UTM projection because this projection can be used to show equatorial areas of the world with minimal distortion in shape and area.

When making community maps at a large scale (small area), you only need to know about one type of projection, that of your base map. If you are looking at another map of the same area and it appears quite different, it could be in another projection. If you want to combine maps from a variety of sources, each on different projections, then you need to understand the principles of projection. It is particularly important to understand projections when working with maps on a computer, because most cartographical software packages allow you to convert a map from one projection to another.

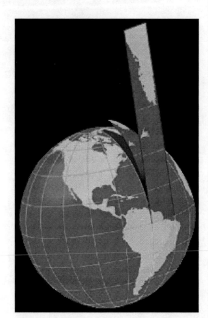

The Mercator projection.

2.3 OVERVIEW OF COMMUNITY MAPPING TECHNIQUES

2.3.1 LANDMARKS: A KEY FOR READING MAPS AND MAKING COMMUNITY MAPS

Whatever method we use to make maps, whether sketch maps or scale maps, the key is noting *landmarks*. We use landmarks as references both to navigate by and to use in drawing maps. A landmark is therefore sometimes called a *reference feature*. In rural areas, landmarks are usually features such as mountain peaks, river junctions, road junctions, large and unique trees, houses, garden huts, or anything else that is distinctive and relatively permanent (for mapping purposes, something that is likely to be around for at least five or ten years).

Whether we are drawing a sketch map on blank paper or drawing on a topographic base map, we start by locating several unmistakable landmarks that we can easily see—for example, a mountain peak or the confluence of two rivers. The locations of everything else can then be measured or estimated in relation to these features. The accuracy of the resulting map is affected by which of various tools and techniques we use to estimate or measure the distance and angle from the landmarks to the features that we are mapping.

2.3.2 AN OVERVIEW OF MAPPING TECHNIQUES

There are several standard techniques for making maps. The table on pp. 32–33 shows the range of techniques commonly used by communities in different parts of the world. Each requires a different set of skills and tools, and a certain amount of time. Each achieves a different level of accuracy. The techniques are listed in order, from the simplest and least accurate to the more technical and more accurate. Often, a few of these techniques are combined in a community mapping project.

These basic techniques are used for surveying and drawing local knowledge on a map. You can read subsection 4.2.3 for information about how to choose the technique or combination of techniques that will be most useful for your community. The community will also have to decide what exactly it wants to draw and how it wants the map to look (see section 4.2 for information on designing thematic maps). There are infinite possibilities. To help you imagine maps of your own community, read further to learn about various types of thematic maps commonly made by rural indigenous communities.

2.4 TYPES OF COMMUNITY THEMATIC MAPS

A number of different types of community thematic maps are described in this section in order to give you a picture of what final community maps might look like. These descriptions should give you some ideas for the design of your mapping project. However, the community that you are working with may not want or need maps of any of these particular themes, and, depending on culture and situation, how they choose to draw their thematic maps may be quite different than described here. Therefore, use this section to gain ideas and to help explain to community members how other communities have drawn their maps, but by all means adapt what you find here to meet the special needs of your community.

Mapping of Boundaries

Boundaries are often the first thing that rural villagers want to draw on a map. In this case, the boundaries that we are talking about are those around customary lands (since legal land-titled boundaries have usually been surveyed and mapped already). The clarification of boundaries is important to everyone: villagers, their neighbours, government officials, logging companies, etc. Traditional village boundaries usually follow natural features, which makes them easy to locate on a good topographic map or air photo. If the boundary is well defined and clear in the minds of the community members, it may be the first information to map, because it creates a framework, a set of outer area limits into which all the other map information fits.

Keep in mind that it is the gravesites, abandoned settlements, and land use that show the local peoples' presence on the land, and that explain why the traditional boundary is where it is. All this other information also helps the people to discuss their boundaries with their neighbours and with the government.

Community boundaries may be well defined according to the local people, but at the same time flexible, not fixed. The boundary may or may not be marked on the ground. It may be an area of shared resource use rather than a fixed boundary. Depict the nature of the boundary as accurately as you can: if it is an area, rather than a line, draw it that way; if it changes from time to time, then draw it as a line and write a note about it, or draw it as several lines with dates on them. Even if the boundary is currently in dispute between neighbouring villages, draw it anyway, and write along the line that it is disputed. Use a special symbol for disputed or uncertain boundaries. Be careful about fixing a boundary that is not traditionally fixed. Too-rigid records of the boundary may cause disputes later. Show the boundary on the map as best you can, and, if necessary, write about the special nature of the boundary or its history in the margin of the map or ona separate piece of paper.

It is critical that boundaries be drawn as accurately as possible. If drawing a sketch, draw the landmarks or reference points that indicate the boundary. Natural landmarks make it clear where the boundary is and the sketch will be useful in discussions (with neighbouring communities, for instance). At a later date, you (or other mappers) can map more accurately with tools such as compass and GPS (explained in chapters 8 and 10), which can be used to determine and draw the exact location of the boundary.

Even if there is a long-standing land dispute, don't wait for it to be resolved. Use a sketch

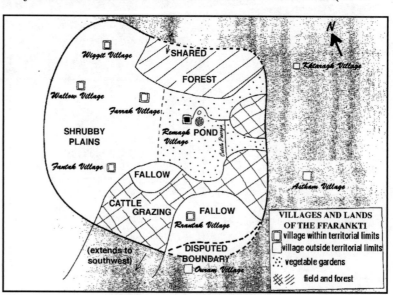

There are many ways, besides a single solid line, to show the nature of a boundary. Boundaries can actually consist of a shared area, or part of the boundary may be disputed.

OVERVIEW OF MAPPING TECHNIQUES
USED IN COMMUNITY-BASED MAPPING

The Mapping Technique	Suitable Purpose for the Mapping Technique	Land Size and Character	Relative Accuracy
Ground sketch (section 5.2) Large pictures of the land, drawn on the ground or on the floor of a house, using any available materials such as plants, rocks, sticks, or household tools.	☞ For internal community discussions, because everyone can see the map, comment and draw ☞ To increase confidence of villagers early in a mapping project	☞ Any size or type of land area	☞ Low
Sketch map (section 5.1) Simple maps drawn free-hand on paper with pencil or coloured pens. They are more permanent and more detailed than a ground map, but are not made to scale.	☞ For internal community discussions ☞ For discussing land boundaries with neighbouring communities. ☞ For planning a mapping survey ☞ For recording local perceptions of the land ☞ For initiating dialogue with government or companies about land-use conflicts	☞ Any size or type of land area	☞ Low
Sketch on topo map (chapter 7) Local knowledge is sketched directly onto a topographic map by locating places in relation to natural landmarks, such as rivers and mountains.	☞ For dialogue with government or companies about land-use conflicts ☞ For land-use planning ☞ For a permanent record of local knowledge and way of life ☞ For making a scale map quickly without surveying	☞ For a large area (over 400 km^2) ☞ For hilly or mountainous terrain	☞ Medium
3-D model (section 7.3) A topographic base map is used as a template to make a three-dimensional model. Pieces of cardboard are cut in the shape of the contour lines and pasted on top of each other. Then the model is finished with plaster and paint.	☞ For internal community discussions about land use ☞ For teaching how to read a topographic map ☞ For making a scale map without surveying, by simply transferring information from the model to the topographic map	☞ For a large area (over 400 km^2) ☞ For hilly or mountainous terrain	☞ Low

THE MAPPING TECHNIQUE	SUITABLE PURPOSE FOR THE MAPPING TECHNIQUE	LAND SIZE AND CHARACTER	RELATIVE ACCURACY
Survey with compass (chapter 9) A compass survey is made by walking along a route, choosing a number of points along it, and using a compass and metre tape to measure the direction and distance from one point to another. A map is constructed with the measurements from the field.	☞ For making accurate maps of small areas ☞ For mapping a village housing area, roads and trails ☞ For making an inventory of forest resources ☞ For making scale maps that have credibility with outside agencies	☞ Especially useful for small areas (under 4 km^2)	☞ High
Survey with a GPS unit (chapter 10) A GPS receiver is carried to places on the land to determine their positions on a coordinate system. Each location can then be found on a topographic map. Alternatively, the locations can be measured onto a coordinate system con·· structed on blank paper	☞ For surveying and mapping large areas quickly ☞ For making scale maps of large areas if you have no topographic map ☞ For locating local places on a standard coordinate system ☞ For surveying boundaries of large areas ☞ For making accurate scale maps that have credibility with outside agencies	☞ Essential for large (over 400 km^2), flat, or hilly areas that are difficult to read on a topographic map	☞ Medium with non-differential GPS ☞ High with differential GPS
Remote sensing (section 6.4) Remote sensing products are photos or images of the land taken with sensors from afar. Air photos are taken from an airplane with a camera. Radar images are also taken from a plane but with radar which can 'sense' through clouds. A satellite image is made by using a computer to process data from satellites to make a picture.	☞ For making scale maps with or without field surveying ☞ For making a scale base map if there is no topographic map ☞ For making scale maps that have credibility with outside agencies ☞ For mapping vegetation, land use, and changes in land use	☞ Use photos for mapping small areas at scales of up to 1:50,000. ☞ Use satellite images for mapping large areas at scales of 1:100,000 or smaller.	☞ Medium to high, depending on the type of remote sensing and the tools used for interpretation

map to help to initiate discussions and to resolve the dispute. Wisely, some village leaders may be reluctant to fix a boundary on their own. A solution is to arrange a meeting of neighbouring village leaders and to draw the boundaries together. If you start with what everyone agrees on, the disagreements might not be that big after all.

Mapping of Culture

A map of cultural features could show abandoned village sites, gravesites, sacred areas, ceremonial places, places to gather forest products used in ceremonies, taboo areas, etc. These kinds of places show the community's history and its unique relationship to the land. On a cultural map you might use point symbols to show specific sites, and area symbols to show larger areas, such as sacred forests.

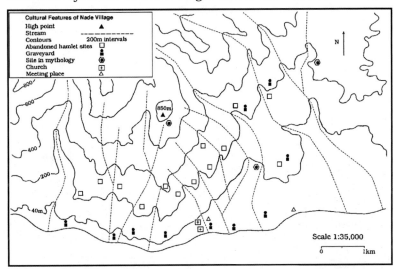

A map that shows local culture and history.

To map cultural places, it is probably not necessary to talk to every individual, because many of the cultural sites are shared by the whole community. Talking to the elders is very important, because many of the cultural features are historical or are related to stories or myth. It is the elders who remember those stories the best. There may be long stories associated with each feature on the cultural map. In that case, it may be easiest to tape-record interviews and later transcribe the tapes to create a booklet to accompany the map (see subsection 11.2.3 for more information on tape-recording). If you are tape-recording, you need to know which site on the map you are talking about. Mark each site on the map with a number and a symbol. Refer to the number during the discussion so that you know which place is being talked about when you listen to the tape later. Carefully label your tapes with the site names and numbers, and the interviewees, so that you can save time finding the right one later.

Mapping of Forest Use

Maps can help document the relationship between the land and the well-being of the community. For example, for many people who live close to the forest the priority information to map is how people use the forest for hunting and gathering forest products. It is usually the traditional forest products that are most threatened by the rapid changes taking place around the community. Many of the subsistence forest products are vital to rural economies, but they are not recognized by the government. In planning for rural development, the government may not see the subsistence value of these products without the help of a community map and explanation.

Document on a map where important forest plants grow. Show where the people of the village traditionally gather different forest products, such as medicinal plants, plants used for household utensils, wood for constructing houses or boats, and green plants and tubers that are used for food. Usually there are so many plant species being used that it is not practical to make a unique symbol for each one. Instead, assign a number code to each species and list the species by the local name (include the scientific or English name if you wish) in the legend. Many plant species grow in specific habitats; for instance, at high elevations in shady locations beside rivers. Try to define these habitats (in as much detail as you think is important) on the map and then arrange the number codes according to which plants grow in each habitat or range of habitats.

The locations where certain animal species are hunted could be drawn on the forest-use map, but it is even better to draw this information on a separate map. Define the habitats in which the animals live and assign a number code for each species. In many regions, animals species migrate according to the seasons. In that case, draw a map for each season (in many places there are four, but the number may be higher or lower). Include the areas where the animals breed or feed or nest.

Over time, ask the elders about each of these plant and animal species, how it is used, and its history. Make a one-page form for each species. Use the same number code on these forms as you used on the maps, so that you can cross-reference this species information to the map. When completed, these forms will contain valuable information. File them systematically in a box or drawer. (See section 11.6 about organizing this supplementary information.)

One advantage of using number codes and local names on the map is that outsiders who might want to exploit the forest products on your land probably won't know what the species are. The forms contain the information that they would need if they wanted

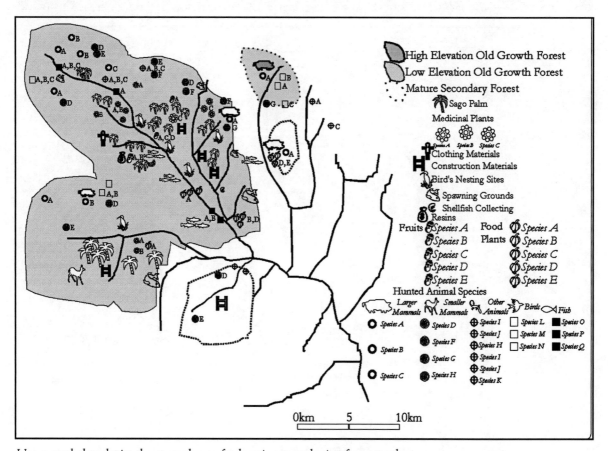

Use a symbol to depict the general area for hunting or gathering forest products.

to take advantage of the local people's knowledge, and it is probably easier to control who sees the forms than the maps.

If the maps are made to scale, it is possible to calculate the size of a forest area. By doing an inventory to count the abundance of forest products (described more in section 11.5) in that area, you can calculate its economic value to the community.

Mapping of Farming and Land Use

Farming and many other types of land use take place in definable areas. Therefore, land use is usually drawn on maps using area symbols or polygons (read the discussions of symbols in subsections 2.2.1 and 4.2.5). In Southeast Asia, a land-use map might include these land-use types: rice fields, fruit-tree groves, rubber plantations, coffee plantations, vegetable gardens, cocoa plantations, *damar* (a resin) forest, rattan plantations, and community forest. Each of these land-use types can be shown as a polygon drawn in a different colour or black-and-white pattern.

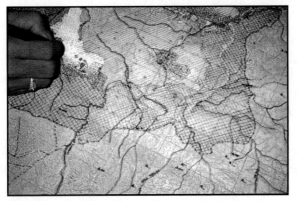

A farming system is often drawn using shaded areas to indicate different crop fields and pastures.

Besides the type of land use, each of these polygons has other information related to it that you might want to include: ownership, the year planted, the combination of species or varieties planted

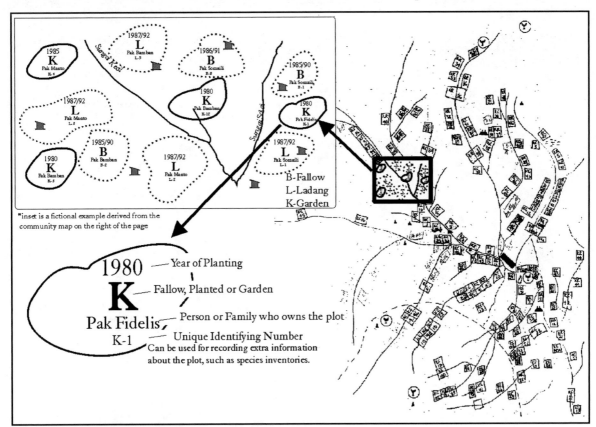

The locations of agricultural land uses are easily defined on a map. Colours or number codes for each field and a list in the legend can show, for example, ownership or year planted or type of crop.

in one land-use area, or the production from each area. It would be difficult to use a symbol such as colour or pattern to show this additional information because it would be confused with the original polygon symbol. But it is possible to use a number or letter code, drawn in the centre of the polygon, to do so. In the legend the code can be explained with the name of the owner, the year planted, and so on.

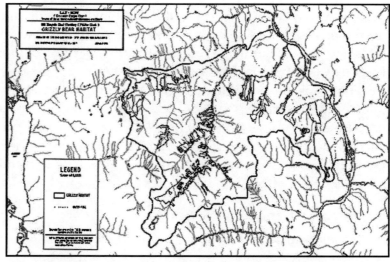

A map showing local people's knowledge of seasonal animal habitats. (Used with the permission of the Strategic Watershed Assessment Team of the Gitksan Nation.)

Land use tends to change over time, and it may be interesting to show these changes on a map. For example, maybe housing is built where vegetable gardens were, rice fields are left fallow, old rattan plantations are harvested and planted with rice. To show changing trends over periods of many years, make a series of maps of the area, for example, in 1981, 1991, and 2001. To show cyclic changes over time, such as a swidden farming cycle, is more complex. Each field or polygon would go through three or four different land-use phases. One year it is planted with rice, two years later with bananas, and fifteen years later with rice again. If the farmers remember this cycle for each field, and often they do, use number codes in the polygons to indicate not just the current type of land use but also the year the land use changes.

The kinds of land uses mentioned above are easily visible on air photos and maybe even on satellite images. Photos, if available, are an excellent way to map general land use. For many places, both current and old photos are available—in that case you can map the land as it was at different times. But the complexity can't be seen.

Mapping Local Ecological Knowledge

A map of local ecological knowledge shows what local people know about where animals live, where different kinds of plants grow, where the soil is best for farming, where fish spawn in the river, which slopes are susceptible to soil erosion, and so on. Local people acquire this kind of information by working an area of land over many years.

Because topographic maps show elevation and slope and river systems, they are very useful for mapping ecological information. Much local knowledge can be gleaned through interviews. It isn't necessary to talk to everyone in the community—it is enough to have a discussion with a small group of people who are known to be good hunters or farmers and knowledgeable about the land.

Elders, of course, have the most experience on the land and must be involved in drawing the map. They may talk about how the animals or plants in the area have changed over their lifetimes. Direct these discussions by asking questions about specific plants and animals, one by one. For example, ask *Where does the wild pig feed? Where does it sleep? Where does it give birth?* Then ask where certain bird species nest. Focus on the plants and animals that are commonly used by local people or are especially important in the culture. These ones will be of most interest to local people and their knowledge of them will be the most profound.

The maps can be made more meaningful by showing greater detail about carefully selected species rather than less detail about a large number of species. Select species that are important to the local people and are good indicators of the health of the various local ecosystems, such as the forest, the rivers, the swamps, and the savannah. Good indicator species are often those that are high on the food chain, such as large mammals (including large herbivores as well), birds of prey, or carnivorous fish.

Mapping of Land-Tenure System

Mapping customary land tenure can be very complex. Indigenous societies often do not have the same concepts of private land ownership as national governments do. Nor do they have strictly communal land ownership concepts. Often there is a defined but flexible combination of both. Generally, certain parts of the community lands are associated with the name of an individual or family by inheritance. Do the best you can to draw the boundaries around these areas of land. Show the name of the 'owner' by writing it on the map or using a number coding system, then in the legend or accompanying notes describe how the person inherited or acquired that land. Also describe the type of control over the land. *Can other people use it? Can others take produce from it or hunt there? Can others borrow the piece of land with permission? Can others buy it?* The amount of detail on this map will depend on the culture and how finely the land is divided.

A more general land-tenure map would not show individual or family ownership, but might just show the kinds of ownership. For example, such a map might show only three areas: an area shared communally according to traditional law; an area for specific commercial use controlled by a community cooperative or business; and an area that is controlled by individual families (without showing the boundaries between the families' holdings.) The type of ownership in each of these three areas would be described in the legend.

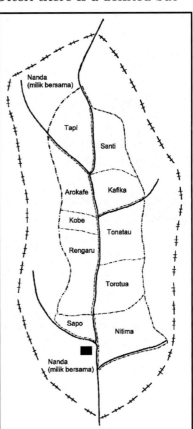

Community land tenure can show individually or collectively owned lands.

3 COMMUNITY PREPARATION AND PLANNING

What you will learn in this section:
- ➢ *How to explain to a local community what maps are and what they are used for*
- ➢ *How to facilitate discussions and decisions about the purpose of mapping*
- ➢ *What the common concerns and risks of making maps are*
- ➢ *How to facilitate discussions to decide what the community wants to map*
- ➢ *What the benefits of a high level of community involvement in making maps are*
- ➢ *How to generate interest for the project within the community*

3.1 INITIATING A COMMUNITY MAPPING PROJECT

In most communities, the idea of mapping is introduced informally at first. Perhaps someone from a neighbouring community is visiting one evening and she talks about her experience of making maps in her own community. Perhaps a student returns home from school with the idea of mapping and begins to talk to some of the elders about it. Perhaps a staff person from a community development organization introduces the idea to the village council. As in these examples, usually the idea is first thought about and accepted by just a few people. Each of those people will have a different idea about maps, depending on their personal experience and interests.

But a community mapping project and the maps that result involve the interests of the whole community. So, at some point early on in the process, the whole community (or as many community members as possible) should meet so that the idea of mapping can be thoroughly introduced and discussed. The key to success in community mapping projects is that all community members feel that they have the opportunity to enter their knowledge and views onto the maps. This process is called ***participatory mapping***. With a high level of participation, the maps will better represent the interests of the whole community. There will be greater motivation to work together to use the maps for the intended purpose. And it will be less likely that conflicts will arise as a result of the maps. How the mapping project is initiated is key to achieving a high level of community participation. When initiating a community mapping project, it is important to have

- ❖ *A thorough and thoughtful introduction to the idea of mapping*
- ❖ *An opportunity for the whole community to make initial decisions about how the mapping project will proceed*

The rest of this chapter outlines how to facilitate a community meeting for the purpose of discussing the idea of mapping and arriving at several major decisions before commencing a mapping project.

Note that this section is written for a facilitator who could either be a person from within the community who likes the idea of mapping, or who might be a staff person from a local non-governmental organization working with the community. At minimum, the facilitator should know the community, have some facilitation skills, and be familiar with maps and mapping (that is, either having had some mapping training already or having completely read through this book).

3.2 HOLDING A COMMUNITY MEETING

Any participatory process will include community meetings. A meeting gives villagers an opportunity to discuss issues together and make decisions in a participatory manner. Well-facilitated meetings allow all sectors in the community to voice their hopes and concerns, and result in decisions that reflect all the voices.

Experience has shown that in the course of a mapping project the community will need to hold two to four meetings. The first community meeting will be to introduce the idea of mapping to the community and to make some initial important decisions, such as the purpose of mapping (see section 3.3). A second meeting of the community (or just of the mapping team) will be to plan the mapping activities (see

The community meets to decide what kind of map (or maps) to draw.

chapter 4). A third meeting (possibly the second for the whole community) will be held after the mapping activities are complete to check and validate the maps before finalizing them (see section 12.4). A fourth meeting might be held to discuss strategies and plans for follow-up and for using the maps.

The first meeting to discuss mapping is especially important. If the meeting is well organized and the participants feel that they have been heard, then they will be confident that the mapping project will represent their interests, and they will be motivated to participate. On the other hand, if these first discussions about the mapping project are disorganized and frustrating, people will assume that the mapping activities will be chaotic also, and the project may never get off the ground. A few tips for making a meeting run smoothly are in the following subsection.

3.2.1 HOW TO FACILITATE A MEETING

Schedule the meeting at a date and time that is convenient for most people. Ask around the village first to check if there are other special events or work activities that might conflict. Then confirm a date. Give people at least one week's advance notice about the meeting. Invite everyone in the community, by word of mouth or with a poster. In whatever manner you use to inform about the meeting, make sure that everyone knows about the purpose of the meeting, the date and time, and the place.

Before the meeting, prepare the materials and the room. You will need large sheets of paper and felt pens to record what people say. You could use a whiteboard or blackboard, but paper is better because it can be rolled up and saved, and the notes transcribed later. Post the paper on the wall or on a stand where everyone can see. Arrange the seating space so that everyone can see the paper and the facilitator and each other. If practical, a large circle is the best for encouraging equal participation in discussions.

Depending on the purpose of the specific meeting, you might also need to prepare materials for presentation: for instance, examples of maps, notes about decisions made at previous meetings, or information about a government decision that affects the community. Write the key information to be talked about, in point form, on large sheets of paper. It is also a good idea to write the meeting's agenda (the topics to be discussed) on a separate sheet of paper and post it.

There are several roles or tasks to be performed in a large meeting. A *facilitator* introduces other speakers and generally maintains the flow of discussion. A *note-taker* writes the proceedings on the board or large sheets of paper. A *timekeeper* keeps track of time and reminds the facilitator about such things as when it is time (as previously arranged) to conclude discussion on the present topic and move on to a different one. In addition, there may be *specialists* who will make presentations about particular subjects, and an *honoured person* in the community may be asked to ceremonially open the meeting. These tasks may each be performed by a different person, or one person could fulfill two or more of these tasks. The organizers of the meeting should meet beforehand with everyone invited to participate in these tasks, so that all agree on the roles, schedule and agenda.

Arrange to open the meeting in a manner appropriate to the local culture. The village headman or an elder might lead a prayer and formally welcome the people. The honorary speaker or the facilitator may state the purpose of the meeting. The facilitator then opens a brief discussion to reach agreement on the meeting agenda so that everyone's expectations are most likely to be met.

This initial meeting will likely be a large group. Large group discussions can be difficult to control. A few individuals may dominate the discussion or take the discussion off on a tangent. The facilitator should make a point of

In a meeting, the facilitator or note-taker writes all the participants' ideas on large sheets of paper.

directly inviting quiet individuals to share their knowledge, as well as posing questions or offering reminders that bring the discussion back on track.

For some discussions, depending on the topic, it helps to break the group into smaller groups. A good size for a small group is six to eight people—enough participants to generate ideas and not too many that any one person feels too shy to participate. In the following sections it is indicated which discussions might be appropriate for small groups and which for large groups.

3.3 DISCUSSING THE MAPPING PROJECT IN A COMMUNITY MEETING

The first meeting's purpose is to introduce and discuss the idea of mapping with the whole community. The objective is to make these important decisions together:

❖ *The community's purpose(s) for making maps*
❖ *Whether or not to make scale maps*
❖ *How to (and who will) control the finished maps and associated information (notes, recording tapes, etc.)*
❖ *Who will take responsibility for organizing the mapping activities*
❖ *What kinds of information and what areas to include when drawing the maps*

The people who are initiating the idea of mapping and who are facilitating the meeting should decide whether or not these discussions and decisions can be made in one meeting. If not, then plan to hold two meetings. This decision depends on the culture, whether the community is accustomed to making decisions together, and how complex the land issues are in this community.

3.3.1 EXPLAIN ABOUT MAPS AND MAPPING

Explain to community members in simple terms what maps are and how they are made. There is no need to explain the technical aspects of maps or to use technical language. Ideally, use the local language. This approach will encourage you to find locally appropriate ways to explain concepts about maps and mapping.

At this stage, community members need to get a picture in their minds about what a community map might look like, what the mapping activities will look like, and how much time and effort will be involved. They need to know, in general terms, the difference between sketch maps and scale maps. Methods for making scale maps can be explained in simple terms, with local examples, in a way that helps people to be confident that, if they want to, they are able to make scale maps. At this stage the villagers need to know just enough to decide whether

A facilitator leads a discussion about what maps are and what is involved in making them.

or not to support the community mapping project and to what degree they personally want to participate in it.

Some of the main points that you might want to explain:

❖ *A map is a picture of the land (show an example if you have one).*

❖ *The community chooses what it wants to emphasize on the picture.*

❖ *Community maps are very different from government-made maps because community maps show local knowledge and concerns about the land.*

❖ *Completing accurate community maps depends on the participation of the community.*

❖ *Villagers are capable of learning to make technical maps.*

❖ *There are a variety of methods for making maps—from simple to complicated, cheap to expensive—and it is the community that decides what method to use.*

3.3.2 DISCUSS THE PURPOSES OF MAPS AND HOW THE MAPS MIGHT BE USED

People who have lived and worked an area of land for generations intuitively know the power of maps as a tool to communicate about the land. Just mention the idea of maps, and communities (particularly those concerned about threats to their land) will be very interested. Therefore, it is usually easy to start a discussion and come up with several ideas about the purpose for making maps from the community point of view. The goal is not to decide on a single purpose. Rather it is an opportunity to discuss, share views, envision—and then write down some of the suggested purposes for the maps. If the purposes are articulated clearly, the people will be more motivated to invest their time and energy in the project, and they will be guided in their decisions about designing and making maps.

Making a map is like telling a story. Before telling a story, you decide how to tell it, depending on to whom you are talking. It is the same before you start to make a map: the first step is to decide what the purpose of the map is and for whom it is being made. People tell stories, or make maps, for different reasons—to record or to communicate different things. A map that is made to fit the community's goals will be more useful in community discussions (or negotiations or presentations) than one made with no purpose or planning.

Use the **brainstorming** technique for this discussion. Take several large sheets of blank paper to write on. Ask the group why they want to make maps. Write the

---

EFFECTIVE WAYS TO INTRODUCE THE IDEA OF MAPPING

☞ *Seek a word in the local language that describes the concept of 'drawing a picture of the land.'*

☞ *Have someone from a neighbouring village describe his/her experience with making maps and using maps.*

☞ *Show examples of maps made by other communities—explain how those communities made the maps, what tools they used, what the participation was like, etc.*

☞ *Show slides or a video of other communities making maps.*

☞ *Involve a big group in making a ground map or some sketch maps.*

☞ *List the local places that the villagers would like to see on a map.*

☞ *Explain the differences between government maps and community maps.*

---

answers as a list on one of the sheets of paper. Write so that the whole group can see it. Record everyone's ideas and save discussion about the ideas for later.

For generating ideas about the purpose of the maps, you could ask questions such as

❖ *Why do we want to make a map of our village/territory?*
❖ *Who do we want to show it to?*
❖ *What are some of our major land issues?*
❖ *Who do we need to communicate with to resolve these issues?*
❖ *What can we use maps for in the short-term?*
❖ *What can we use maps for in the long-term?*
❖ *Do we have a defined, concrete purpose for the maps (for example, to present in a court case)?*
❖ *Do we also have intangible purposes, such as bringing the members of the community together to consider the community's collective future?*

After the list is complete, discuss its contents. Rewrite the list, combining ideas that are similar and using specific terms to restate ideas that were first stated in general terms. It is not necessary to have a single purpose for mapping, nor a hierarchy of priorities. In fact, most communities have multiple purposes of equal priority. What is important is that the purposes are discussed in depth and articulated clearly, by and for the community.

3.3.3   MAKE A LIST OF WHAT INFORMATION TO MAP

Take a new sheet of paper (or clear the whiteboard after first having recorded the list of purposes on paper) and do another brainstorming session. This time, make a list of all the kinds of local knowledge and places and activities that the community

---

SOME PURPOSES FOR COMMUNITY MAPS

☞ *To open a dialogue about land-use conflicts with government departments, logging companies, plantation companies, mining companies, or ranchers*

☞ *To resolve a boundary dispute with neighbouring communities*

☞ *To prevent boundary conflicts with neighbours*

☞ *To help plan land use, land protection and local economic development*

☞ *To raise community awareness about the land and the local environment*

☞ *To increase villagers' confidence in their leaders' ability to manage and protect their customary lands*

☞ *As a tool for community organization*

☞ *As a database of local knowledge to strengthen the community's bargaining position*

☞ *As an advocacy tool for litigation, negotiation, or lobbying*

☞ *As a record of the community's history, traditional law, traditional land-use systems, and way of life*

☞ *As a tool to revitalize the local culture*

☞ *To help educate the children about the community's traditional land laws*

☞ *To substantiate the community's history and stability in this location*

---

members want to see on a map. List everything and anything. As an example, your list might include rivers, roads, hunting trails, rice fields, fishing places, houses, hunting camps, gravesites, ancient battlegrounds, boundary markers, taboo forest, and so on.

This discussion is also good to do in small groups according to sector—for example, elders, women, youth, farmers, and hunters. (Of course, some people would fit into several groups, so let it be their decision to choose whichever group most interests them.) Each group is asked to make a list. When finished, the groups join together in the big group again and combine the lists. Rewrite the combined list, eliminating duplications and clarifying items that aren't clear.

Break into small groups and make a list of what to draw on the map.

If you like, you can discuss what information you need for each particular mapping purpose. Take a look at the example from a community in Thailand (p. 46) to help think about what information is relevant for the community that you are working with.

At this stage, listing what to draw helps people to imagine the community maps that they want to make. Later, in subsection 4.2.1, this list of information will be used to plan what kinds of thematic maps the community will make.

3.3.4   DISCUSS AND DECIDE WHAT AREA TO MAP

Besides what kinds of information to draw, you should also discuss what geographical area to draw. *Do you want to map the whole traditional territory, or, because of urgency, do you want to map just one threatened part of the territory now? Do you want to map just one village, or do you want to map a cluster of villages together on one map? Do you want to*

ACKNOWLEDGE WOMEN'S KNOWLEDGE OF THE LAND

When first introduced to mapping, many villagers believe that men have the most knowledge to contribute, because men travel farther on the land for hunting and gathering forest products than do women. Men may seem to have more to say on boundary issues and forest use. Men's role in land use often appears more important because it shows the outer extent, the biggest area of land use. Also, the family lands may be in the name of the husband of a family, and so he does the talking.

However, women do in fact have much to contribute through their detailed knowledge of certain areas of the territory and about different subjects than the men. Although women's knowledge might not show the extent of the land base used by the community, it may better show the intensity of land use. In many cases, a map of forest products would not be complete without the women's knowledge of medicinal plants and other products gathered by women, and where to find them. Women's knowledge is important to include on the maps, because without it you have only half a picture of the community.

REASONS TO MAKE A MAP: AN EXAMPLE FROM THAILAND

Members of one village in Thailand had a meeting and came up with these reasons for making a map:

WHY DO WE WANT TO MAKE A MAP?	WHAT INFORMATION DO WE NEED TO DRAW?
☞ To demonstrate to foresters and agriculture officials that our land-use system is more complex than they think and that it sustains us	☞ Location of protected community forest area ☞ Forest products ☞ Animal grazing areas ☞ Our seasonal use of the forest ☞ Locally designated agricultural land ☞ Rice field rotation ☞ Location of fruit tree gardens
☞ To feel security on the land	☞ All of the above
☞ To show the government that we live sustainably in a stable location	☞ Village gate ☞ Our traditional boundaries
☞ To record our spiritual beliefs about the land, to tell outsiders about our culture and to teach our children	☞ Sacred watershed ☞ Sacred pond ☞ Burial ground ☞ Swing place, with bamboo swing poles ☞ Community ground ☞ Sacred trees ☞ Places to collect carving wood
☞ To prevent land disputes between our villages and other villages	☞ Our village boundaries ☞ Family land boundaries ☞ Traditional land-use rights system ☞ Landmarks of historical significance

map within the modern administrative boundary or within the traditional boundary of the village?

The group could draw a sketch map to help discuss these questions. Also, refer back to the list of purposes to help the group come to a decision. As an example, if a priority purpose is to revive the local culture or assert traditional ownership of the land, then perhaps two or three administratively separate villages may choose to work together to draw one map, because traditionally they shared the same territory.

3.3.5   DISCUSS RISKS AND CONCERNS

During the discussion about whether or not the community wants to make maps, some people in the community will express concerns. Indeed, although there are many good reasons for making maps, there are risks too. Risks are usually associated with how the maps may be used (or misused) rather than with the maps themselves. By considering the risks in advance, you can avoid problems. For the community to consider, here are some common concerns that have been expressed by various communities, and some ideas about how to address them:

Local knowledge and oral history is sacred and alive and can't be correctly translated onto a map. Much of local knowledge and traditions about the land are communicated from generation to generation through stories and songs and just talking and living on the land. Don't expect maps to replicate this living transmission of knowledge. But walking on the land for the purpose of mapping can revive the storytelling. Of course there is also risk of losing the meaning of local knowledge when trying to fix it onto a topographic/scientific frame. Therefore, it is important to pay attention to the community process, the ground maps, and the sketch maps.

Maps of forest products might help outsiders to find and over-exploit resources (fruit trees, fishing grounds, bird's nests, etc.) that are important to the community. Be sure to consider the advantages and disadvantages of putting such features on any map that will be used outside the community. You could make one set of maps that are tightly controlled by the community, and draw more generalized maps for public viewing. Alternatively, draw only the forest products not likely to have commercial value.

Maps of cultural and historical areas (showing, for instance where to find carvings, sculptures, or culturally modified trees) might lead outsiders to plunder them. These items need to be considered in the same way as the resources mentioned above. Weigh whether these areas are more threatened or more protected by showing their locations on a map. Outsiders may come to visit these sites if they know where to find them, and may not respect them the way a villager would. However, if a logging company is willing to avoid the areas during its operations, then a map is needed to show where they are. The risk of showing these features on a map is diminished if the community is confident of its negotiating position.

Maps of their land can raise the expectations of villagers too high. Maps alone will not protect the land. It is villagers working together using the maps that can help to protect the land. From early in the project, have frequent discussions about strategies to use the maps.

Fixing a boundary line on a map can aggravate boundary disputes. There may be disagreements between individuals or between groups about who has claim to certain land or territory. Such disputes can be aggravated by trying to draw a line on a map. Try drawing a sketch map first to help discuss and resolve the boundary location. Or don't draw a boundary *line*—draw a *shared area* at the boundary. Indicate clearly on the map that a boundary is in the process of being negotiated, or that several families claim ownership of a certain area. Be careful not to help entrench positions. Postpone the resolution of disagreements until more information has been collected. Start by recording history and land use instead. Use the process of making maps to help people to talk together and reach agreement.

Maps can be misleadingly definite. When people see a line on a map, they tend to believe it. This power of the image can be both good and bad. It is what makes maps so useful as a good communication tool for communities who feel left out of land-use decisions. On the other hand, it can create conflicts when a stakeholder thinks that a line on the map is fixed and not negotiable. Indicate clearly on the map if it is still in progress—a 'working' or draft version still being added to and revised—so that viewers will understand that it is not a final map and that they can still suggest changes.

It will take years to make complete and accurate maps of the community way of life, and an incomplete map will be easily misinterpreted or used against the community. Write a qualifier in the legend of the map that states that the map is in progress.

3.3.6 DISCUSS AND DECIDE WHO WILL CONTROL THE FINISHED MAPS, AND HOW

A key to resolving many of the concerns listed above is to ensure community control of the finished maps. *Who outside the community will be allowed to see the maps? Who, inside or outside the community, will be allowed to have copies of the maps? Who in the community will be responsible for enforcing these decisions about how the maps are distributed? Where will the maps be stored in the community?*

The answers to these questions will depend on the kind of decision-making or leadership structure that exists. Perhaps a village chief or a village council will control the maps. The maps will need to be more tightly or less tightly controlled depending on how vulnerable the map information is to misuse. If the maps are being used for nego-

tiations or legal court cases, then they need to be tightly controlled in the community. Or, if people are concerned that information about the location of medicinal plants or gold or other resources could be exploited, then the distribution of maps containing this information would also need to be tightly controlled. Often, a community will decide to closely guard certain thematic maps and not others.

The first community meeting to discuss mapping.

IMPORTANT COMMUNITY DECISIONS ABOUT A MAPPING PROJECT

☞ *What are our purposes for making maps?*

☞ *What information and area do we want to map?*

☞ *Do we want to make sketch maps only at this time, or do we want to undertake a technical mapping project?*

☞ *Who in our community will be responsible for facilitating the mapping work until it is finished?*

☞ *Who will control the use of the maps when finished and how?*

3.3.7 DISCUSS AND DECIDE WHETHER TO MAKE SCALE MAPS

At some point the community needs to decide or confirm together whether it wants to make a scale map, or whether a sketch map is sufficient for its purpose. Discuss the advantages and disadvantages of making scale maps. Scale maps have more credibility with outsiders, but they require more time, technical skill, training and financial resources to make than sketch maps. Some people in the community who have no formal education might feel inhibited to participate in technical mapping activities.

If so, and participation is a priority, then sketch maps might adequately serve the community's purpose at this time. Or the community might decide that making scale maps is too costly in terms of time and effort. Of course, some methods for making scale maps are more costly than others. Make sure that community decision-makers are aware of all the options. Read more about the different mapping methods in section 2.3 to help you facilitate this discussion.

Members of a community mapping team organize the mapping activities and encourage participation.

3.3.8   DISCUSS AND ESTABLISH A MAPPING TEAM

In participatory mapping we hope that everyone in the community gets involved in some way, but we don't expect that everyone participates equally in every activity. A mapping team should be selected to organize the mapping activities, to encourage participation, and to take responsibility for finishing the maps. This mapping team will accept the responsibility of fulfilling the mandate given to them by the community as a result of the decisions made at this initial community meeting.

The mapping team could be any size from two to six people and should include active elders who know the land and local language and culture. If the community decides to make scale maps, then the team should also include youth with a high-school education. The members of the team should be enthusiastic, keen to learn new skills, and be respected and trusted by the community.

Match up the wisdom and knowledge of the land that the elders carry with the school-learned skills and enthusiasm of the young people.

3.3.9   CONFIRM COMMITMENT TO COMMUNITY PARTICIPATION

Community participation is essential for an effective mapping process. Involving many people from the village means that the mapping process may take longer than if it is done by a team of experts, but the results, in the long run, will be better. The key to achieving optimal community participation is to make it a priority from the beginning. Discuss the issue of community participation at the initial meeting and confirm the community's commitment to a participatory approach. To help you facilitate this discussion, read in the box below why community participation is important.

WHY COMMUNITY PARTICIPATION IS IMPORTANT

Full involvement of local people will produce better, more complete maps. *It might be a special team of young people from the village that do the surveying and drawing, but they probably know only part of the history. All of the knowledge and stories about the land must be drawn on the maps to show the whole picture. The maps will be a better picture of the community land if a wide range of people are represented, including both women and men; young and old people from different families; and people who are recognized experts on particular subjects, such as hunting, fishing, healing, food-gathering, etc. The members of the community alone have the unique knowledge with which to draw an accurate map of their history, their land use, and their culture.*

Making maps creates opportunity for community members to learn more about their own land. *Youth in the community now spend most of their time at school and don't have the opportunity to learn about the land. Adults focus on activities on their own farm or in their own business and don't get around on the land. By being involved in a mapping project, these people are motivated to walk on the land and learn about the land from elders. If the whole community knows the land, its members can be much stronger in fighting to protect it.*

Making maps creates an opportunity for different members of the village to work and to learn together. *Some people have an especially rich knowledge of the history and culture of the people—or detailed memories about how things used to be, and how they have changed. Other community members are particularly skilled in drawing maps or organizing mapping projects. Still others are particularly good at securing the community's needs in negotiations with government agencies. By working together, villagers can enrich their knowledge at the same time as they share their skills for the benefit of the whole community.*

If all the community members participate, then they will better understand the process of making maps, so that they can trust that the maps are a true record of their knowledge. *Afterward they will have more confidence and credibility when they discuss the maps in meetings with NGO people or with government or company officials.*

Participating in making maps can be a starting point or tool to empower the community to be active in making its own projects and plans. *The maps can be used by the community for land-use planning. Use the process of researching and making maps to encourage the community members to use their own vision, initiative, and resources to help them to direct their own future.*

PLANNING TO MAKE MAPS

What you will learn in this section:
- ➤ *How to design a mapping method appropriate to the community*
- ➤ *How to determine what thematic maps the community needs to draw*
- ➤ *How to estimate what scale is appropriate to use*
- ➤ *How to weigh various factors in order to choose an appropriate mapping technique*
- ➤ *What symbols and language to use for field mapping*
- ➤ *What materials are necessary for different mapping techniques*
- ➤ *How to arrange a training session for mapping*

4.1 PLANNING TO MAKE COMMUNITY MAPS

This chapter is about doing the detailed planning necessary in order for community members to make scale maps to meet their own needs. If in the first community meeting it was decided not to make scale maps, just sketch maps, then skip ahead and read further in chapters 5 and 11 about how to gather local knowledge, and then adapt these simple techniques for the local community. If a sketch map is what the community needs, then it doesn't take any planning; you can start making a map right away. However, making scale maps does require some planning. This chapter will give you ideas about what things to consider in order to plan a mapping method and get into action.

4.1.1 WHO DOES THE PLANNING?

It is good, but not necessary, to do the detailed planning with participation from the whole community. Of course, if the maps are designed by all the community members together, the maps will best reflect how they live on the land and will emphasize what is important to them. Community members will then more easily understand and use the final maps, and they will take pride in them. On the other hand, making certain decisions, such as what scale to use, may require technical understanding that not all community members want to master. Planning can be done effectively by the mapping team and a few key people in the community, such as respected elders. The initiators of the project can assess the capacity and dynamics in the community—as well as the enthusiasm at the first community meeting—to decide whether planning should be done by a large group or by just the mapping team that was selected by the community. When the community selected a mapping team, it indicated its trust in the team to make the detailed decisions about implementing the work. The team was given the mandate to decide *what* to draw on the maps and *how* to survey and draw the maps.

Regardless of whether many people in the community want to be involved in this more detailed planning stage or whether it will involve just the mapping team, the first step is to organize another meeting. In order to respect the community decision-making process, the planning decisions to be made in this meeting must accord with the decisions made at the initial community meeting (section 3.3). To remind everyone about those earlier discussions and decisions, write the key points on large, clean sheets of paper and post them at the front of the meeting room.

4.1.2   USING A SKETCH MAP FOR PLANNING TO MAKE SCALE MAPS

It will help the discussion and planning to start by drawing sketch maps of the area that the people intend to map (see chapter 5). If the community has already drawn sketch maps, then bring the maps to the meeting and display them for all to see. For planning purposes, the sketch maps should have on them the major landmarks and their local names—for example, mountains, rivers, lakes, major roads, and commonly used trails. They should also show the outer boundary of the traditional territory (unless the boundary is under dispute or is otherwise difficult to draw).

Use a sketch map to plan a survey.

4.2   DECIDING ON A MAPPING METHOD

Many indigenous peoples around the world have made scale maps describing their land use and history. They have developed various ways of doing it, depending on the purpose of the maps, their culture, the characteristics of their land, and the skills, equipment and time that they had available. There is no expert to tell you exactly how to do it. Every village, every culture, is unique. This section will help the community or mapping team to make the planning decisions necessary to get started.

4.2.1   CHOOSE WHAT THEMATIC MAPS TO DRAW

At the first meeting, the community made a list of all the information to be drawn on the map. Even after the list has been refined and shortened, it probably still lists too much to draw on one map. Try grouping the items into subjects such as *history, land use, forest products* and *culture*, or whatever works for the community's needs. Each of these subjects will have its own thematic map. In the end, for example, you might decide to make three thematic maps for the area: *land use, ecology,* and *culture.* The land-use map might show swiddens (areas of shifting cultivation), fruit trees, and wet rice all on one map. The ecological map might show areas of different kinds of forest. The cultural map might show the gravesites and abandoned settlement sites. By separating the different kinds of information about the community onto a series of thematic maps, you avoid ending up with a map that is too cluttered and confusing. Review again in section 2.4 the scenarios involving different types of thematic maps.

If you have a lot of information to show, you could divide each of these subject areas into more specific categories. For example, you could make three land-use maps: *forest products, swidden rotation,* and *permanent agriculture.* Three separate ecological maps

A SUGGESTION FOR ORGANIZING DIFFERENT KINDS OF INFORMATION ON THEMATIC MAPS

BOUNDARIES	CULTURE	FOREST PLANTS	LAND USE
Family land	Gravesites	Medicinal plants	Rice gardens
Village land	Rock carvings	Food plants	Rattan plantations
Tribe land	Sacred trees	Construction wood	Fruit tree orchards
	Historical sites	Weaving material	Coffee plantation

(If you find that different categories work better for you, then use them instead.)

could show *soil types, slope angle* and *vegetation cover*. Three cultural maps could depict *gravesites and sacred sites, village boundaries,* and *land tenure* within the village boundaries.

Time and space are also ways to categorize information. You might want to show the history of swidden rotations. *Where did each family open their first swidden from original forest in the area? Where did they make swiddens after that?* By making a map for one time period and another map for another time period, you can show how the land has changed. You could make **before-and-after** maps to show what the village land looked like ten years ago, and what it looks like today. And, ten years from now, you can look back at the map that you are making today and see what else has changed.

4.2.2   DECIDE WHAT SCALE TO USE

In choosing the scale, you should be guided by the purpose of the map, making sure that the essential features can be depicted clearly, while keeping the physical size of the map manageable. In choosing the scale of the map, you have three considerations:
   ❖ *The size of the paper that you want to use*
   ❖ *The size of the land area that you want to map*
   ❖ *How much detail you want to draw*
Having decided what kinds of information you want to draw, you now have an idea of how much detail you will need to draw on the maps. The scale (see subsection 2.2.2) sets a limit on the amount of information that can be included and how it can be shown. Assume that the smallest area that you can easily draw onto a map and still create a recognizable pattern is probably about 5 mm across. If, for instance, the map scale is 1:50,000, that 5 mm on the map represents a real-life area 250 m across—any feature that is smaller, if it is important to show it, would require a point symbol instead.

Though we realize that all land maps are smaller than the real size of things on the ground, too small a size will limit the number of features that you can show clearly. Conversely, the larger the scale, the more detail you can show, because everything is drawn bigger. But too large a scale creates handling problems—for example, the challenge of folding and unfolding huge map sheets in rain and wind. You have to consider how to carry the field maps and how you can draw on them in rain or sun, and on uneven ground. It is possible to cut a large map into sections for easier handling, but doing so may be awkward or confusing to work with. (If you do divide the map, be sure to coordinate the information on adjoining sections! It helps to first survey the boundaries between adjacent sections before mapping the interior of each section.)

Having sketched the area that you want to map, you can estimate the size of the area. Estimate the size of the area in kilometres (or metres) wide and long. Then, since at this point you are working with only estimates anyway, try using the table below to choose an appropriate scale. For this table we assume a standard paper size of 60 cm × 80 cm. If you want to use a larger size paper, then choose a scale slightly larger than indicated in the table.

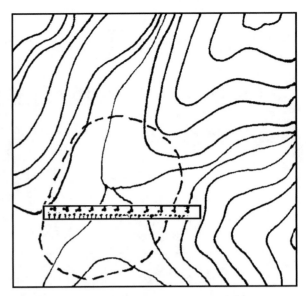

Sketch the area on a topographic map and then measure the length and width to calculate the scale needed.

CHOOSING A MAP SCALE

(Assuming that the paper size is 60 cm × 80 cm and the map size is 50 cm × 50 cm)

RELATIVE SCALE	SCALE	LAND SIZE (LENGTH AND WIDTH)	AMOUNT OF DETAIL
Very large scale	1:500 to 1:4000	250 m to 2 km	Very high
Large scale	1:5000 to 1:20,000	2.5 km to 10 km	High
Medium scale	1:25,000 to 1:150,000	12.5 km to 75 km	Medium
Small scale	1:200,000 to 1:500,000	100 km to 250 km	Low
Very small scale	over 1:500,000	over 250 km	Very low

At this point, in order to choose a mapping technique, you really only need to know the approximate scale at which to make the map. If you already have a base map and you want to calculate the scale more precisely, use the equation in the chapter about preparing base maps (subsection 6.2.2).

4.2.3   CHOOSE A MAPPING TECHNIQUE

There are several factors to consider in deciding which mapping technique to use:

❖ *The size of the land area and its characteristics*
❖ *The time available*
❖ *The published map information that is available*
❖ *The level of accuracy required*
❖ *The resources available (people, skills, tools)*

Think about these factors and follow the decision tree on the next page to help you decide which mapping method is most appropriate for the situation. Refer back to the overview of mapping techniques in subsection 2.3.2 for a summary of each technique and the purpose for which it is most suited.

With these factors in mind, choose a mapping technique according to the purpose of the map. The purpose should already have been decided upon by the community in the first meeting. If you want a map to help you in a legal case or in negotiations with government, then you would choose technological methods to make a reasonably accurate scale map. If you primarily want the map for internal community discussions, then the map might be more meaningful as an informal, colourful sketch map with many picture symbols and local place names that people of the village can relate to. Although the content of both these kinds of maps is traditional knowledge, the final form of the maps is different. (Recall the different kinds of maps discussed in section 2.4).

The fastest way to make scale maps is by drawing directly onto base maps made from existing maps or air photos (or the products of other types of remote sensing). This technique is useful if you need maps urgently, for instance if you have an immediate opportunity for dialogue with the government about a particular threat to the land. A great deal of local knowledge can be drawn directly onto base maps, with reasonable accuracy, without doing any time-consuming field surveying. If a little time is taken to teach local people how to interpret the existing base maps, this technique can also be highly participatory. The effectiveness of drawing straight onto base maps depends on the characteristics of the land. It is most effective if the shape of land (and other landmarks) are evident and well defined, as it is in mountainous areas. If the land is flat and featureless, then you may have no choice but to map the area with field surveys.

Another situation where you might want to make scale maps by drawing directly on base maps is if the area to be mapped is very large (over about 400 km^2) and you don't have the time to walk and survey around it. In this case you could draw the outer boundary of the territory on a base map (you might require more than one map sheet for the territory). You immediately have a useful scale map. Then, if you have the time and equipment, select certain sites or areas to survey in the field. For example, if the location of parts of the boundary are not evident on the base map, then you may want to use a GPS unit to survey them in. If there are features (for example, rice fields) for which you want to calculate precise areas, then you can survey around them with a compass and metre tape.

In order to choose a field surveying technique, estimate the size of the land area that you want to map, then weigh the amount of time and accuracy required for each method. A compass survey is much more accurate than a survey with a non-differential GPS unit (see subsection 10.1.4 for a definition), because the location of a GPS coordinate can be off by as much as 30 m. But, for a large land area, surveying with the GPS is much faster than with a compass. And, because a large area is mapped at a medium or small scale, the 30-m error will represent only a few millimetres on the map. So, for a large area (over about 25 km^2), the accuracy of GPS data (even if uncorrected) is quite adequate. For a small area, a compass survey is the best option because it is feasible in terms of time and is much more accurate than the GPS.

A mapping project is sometimes planned for a very specific purpose, such as negotiating compensation for a small area of forest (say, 500 m × 500 m) that was bulldozed by accident by a logging company. In such a case you would need only one mapping

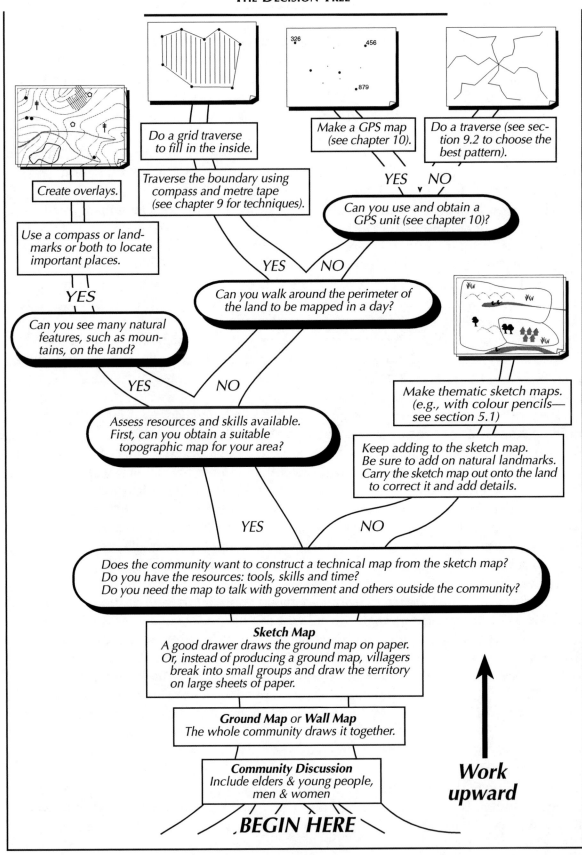

Create overlays.

Do a grid traverse
to fill in the inside.

Make a GPS map
(see chapter 10).

Do a traverse (see sec-
tion 9.2 to choose the
best pattern).

Use a compass or land-
marks or both to locate
important places.

Traverse the boundary using
compass and metre tape
(see chapter 9 for techniques).

YES   NO

Can you use and obtain a
GPS unit (see chapter 10)?

YES

YES   NO

Can you walk around the perimeter of
the land to be mapped in a day?

Can you see many natural
features, such as moun-
tains, on the land?

YES   NO

Make thematic sketch maps.
(e.g., with colour pencils—
see section 5.1)

Assess resources and skills available.
First, can you obtain a suitable
topographic map for your area?

Keep adding to the sketch map.
Be sure to add on natural landmarks.
Carry the sketch map out onto the land
to correct it and add details.

YES   NO

Does the community want to construct a technical map from the sketch map?
Do you have the resources: tools, skills and time?
Do you need the map to talk with government and others outside the community?

Sketch Map
A good drawer draws the ground map on paper.
Or, instead of producing a ground map, villagers
break into small groups and draw the territory
on large sheets of paper.

Ground Map or **Wall Map**
The whole community draws it together.

Community Discussion
Include elders & young people,
men & women

BEGIN HERE

**Work
upward**

technique. Survey by compass around the perimeter of the area in order to accurately calculate the number of hectares. Estimate the value of a hectare of that type of forest (perhaps by making a forest inventory, as is introduced in section 11.5) and you can calculate a fair compensation.

More commonly, communities are interested in mapping their whole territory, and for multiple purposes: for dialogue with government decision-makers, to create a record for their grandchildren, for land-use planning, etc. To make such a multi-purpose map, you would need to combine several mapping techniques. If you want to produce a record of the whole territory and a tool for dialogue, you can use a GPS receiver and a topographic map for locating and drawing the boundary. In order to plan a housing development, use a compass survey along the main village roads. For mapping forests, interview hunters and sketch directly on a topographic map.

4.2.4 DISCUSS TECHNIQUES FOR PARTICIPATORY GATHERING OF LOCAL KNOWLEDGE

Take the opportunity at this meeting to discuss what techniques to use for participatory gathering of local knowledge. Look at chapter 11 for some ideas. Use several techniques or activities in order to discover as much information as possible and to maximize the opportunity to cross-check the information that emerges. Depending on the social dynamics of the community, some techniques will be more appropriate than others.

GETTING THE COMMUNITY INTERESTED AND INVOLVED

☞ **Use the local language and local names on the maps.** *That way, community members will feel that the maps are relevant to them and feel more comfortable participating in the process.*

☞ **Make the mapping activity accessible to everyone.** *Post an extra copy of any map that you make in a place where any community member can come and look at it at any time. Provide paper and pencil with an invitation to leave comments.*

☞ **Show people that mapping does not have to be complex.** *A good way to start is to help the elders to draw sketches of the land. Then walk on the land with the elders and ask them to point out special places. Carry a map and draw as they talk. Sometimes elders remember details about the land only in songs or stories. Listen to the song or story and write it down. Then ask questions to guide you to the place in the song. Then draw it on the map. Doing so can help to keep the traditional knowledge alive. It is important to record all the information before the elders die and people forget.*

☞ **Provide training.** *With just a little training, local people are eager to use the maps to record additional information on their own, or they might use the maps to assist in resource monitoring. Some aspects of map-making are simple, some are complex. Many villagers intuitively know the benefits of a map. It is not difficult to engage their interest.*

☞ **Start by focusing on topics of special interest to particular community members.** *Ask hunters where they hunt, farmers where they farm, carvers where they find their wood, ceremonial leaders where the sacred places are, healers where to find medicinal plants. Where a location has traditionally been kept secret, discuss with the person the advantages and disadvantages of recording the site on the map, and be sure to respect his or her feelings on the subject.*

Integrate activities for gathering local knowledge with the mapping activities. In a typical mapping project in Indonesia, small groups of villagers do field surveying during the day, then gather in a central house at night to have discussions around a three-dimensional model, or to draw a seasonal calendar, or to record a story from an elder about one of the sites they surveyed that day, or to sketch additional information from memory onto the base map. While they are sharing local knowledge in this way during the evening, they realize where or what they want to survey the next day.

Brainstorm ideas about how the mapping team can facilitate the various activities in a way that motivates villagers to participate and to share their knowledge. How can the team make the mapping and documentation interesting and fun?

4.2.5 DESIGN SYMBOLS TO USE FOR FIELD MAPPING

Maps are, by definition, graphic. After all, it would be impossible to write everything on the map in words. There would not be enough space on the paper and it would be confusing to read. Before starting the mapping project, decide on symbols to use in the field. In participatory mapping, the most important thing is that the symbols are attractive and interesting and simple enough to be drawn by anyone in the community. And, from a cartographic perspective, the symbols need to be distinctive enough

---

A CHECKLIST FOR THE COMMUNITY ON ISSUES RELATING TO INTELLECTUAL PROPERTY RIGHTS AND PROTECTING MAPS THAT SHOW LOCAL KNOWLEDGE

☞ Are there community institutions that are well established, respected, and representative of all community sectors?

☞ Has the community (or a committee of it) made guidelines and policies for controlling and using maps of local knowledge?

☞ Are there customary (adat in Indonesian) restrictions on the use and distribution of certain types of information (for example maps of sacred areas or a valuable resource)?

☞ How and by whom will the mapping of local knowledge be conducted?

☞ Are there customary laws to obey in the process of documenting local knowledge on maps (ceremonies to perform, compensation to pay)?

☞ Are community members clear about the expected benefits and impacts of making maps?

☞ What is the community's role in verification and review of all documents?

☞ How will the maps and related documents be made available to the community?

☞ Do sponsors or financial donors claim any rights over the information? If so, how will this situation be dealt with?

☞ Is there a well-worded copyright statement?

☞ Is there a location where maps can be stored safe from fire, flood, or theft?

☞ Is there a second location where copies of the maps can be stored safe from fire, flood, and theft?

that they cannot be confused with each other. However, try as much as possible to use symbols (such as pictures, or coloured lines and areas) that can be recognized even without a legend to represent what you want to show.

Design the symbols with a group of villagers. You already have a list of information that the community wants to map. Write that list on a large sheet of paper and draw a symbol beside each item on the list.

For specific places on the land (such as a house, a sacred rock, or a small graveyard), use point symbols. Point symbols can be simple pictures, or they can be abstract shapes, such as circles and squares. Point symbols could also be numbers or letters. If you choose to use picture symbols, keep them very simple so that they can be drawn exactly the same each time—so that a picture of a bird is not mistaken for a pig or a house. Remember too, that simple pictures are faster to draw, and easier to draw small, than are complicated ones.

Local people will intuitively understand lines as symbols representing roads, rivers, and territorial boundaries. Although there are an infinite number of patterns of line symbols that you can create if you need to, you might want to begin with the ones that are traditionally used for mapping. For instance, to show dirt roads or trails with a dashed line is standard. A paved road might be a thicker or double dashed line, or a solid one. Rivers are always a solid line, unless they are seasonal only. Boundaries could be a string of x's or o's, or a pattern that alternates dots and dashes. If you have several different types of boundaries on one map, be creative and think of different types of line

Have the community members work together to design the symbols, using the local language, so that the symbols are meaningful to them.

symbols that are clear and easy to differentiate. When making field maps, dashed lines can be confusing, and it is best to use solid coloured lines. Colour makes line symbols even easier to recognize.

For participatory mapping, colour is the best way to identify different areas (polygons) on the rough maps. Colouring polygons is fast and fun. A coloured map is more interesting to look at and easier to understand than a black-and-white one, especially for elders with poor eyesight and for large groups crowded around a single map sheet. Colour pens or pencils are not as easy to erase as regular pencil, but, if necessary, you can make corrections with 'Wite-Out' or similar correction fluid on the field maps.

Be creative in designing symbols for the field maps. Don't worry about using standard symbols. Later, when you redraw the maps as final maps, you will want to use templates, stick-on symbols, or a computer for a more consistent and neater appearance (see subsection 12.2.2). You may use some of the same symbols as in the field, but likely you will change other symbols, for example the coloured polygons.

4.2.6 CHOOSE THE LANGUAGE TO WORK IN

What language should you use for questioning and for recording information? Use the language that villagers most commonly use for talking about the land. Using the local language helps to validate the local knowledge, which is important if the community is to feel that the maps are relevant to them. It will make it easier to sustain the community's interest and involvement and build their confidence in making maps and discussing maps. Using the local language also helps to improve the accuracy of the map by eliminating confusion and misinterpretation.

Before beginning to map, it may be useful to develop a dictionary of the terms that will be needed throughout the mapping project, especially words for describing the land. Frequently, villagers will describe a location using local terms for geographic features such as ridges, valleys, headwaters, or river junctions. You will often find that local terms won't translate directly into the national language. Local languages may express greater detail within certain categories of information. Much local knowledge is transmitted in the form of stories and legends, using metaphors and unique terms. By trying to translate literally and directly, you can miss information. When translating, think first of the whole meaning of the term or concept. If you make a dictionary, consider making it a *picture dictionary*. It might be useful to organize the dictionary by topic rather than alphabetically; for instance, land shapes in one section, crop names in another, types of land tenure in a third, etc. Finally, always be ready to correct the dictionary as you develop new understanding.

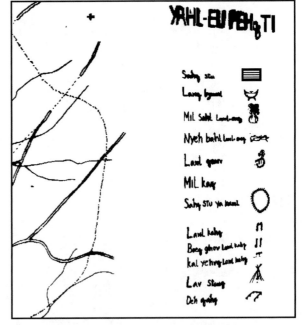

For internal community use, make the map in the local language. It affirms the local language and it could help to protect the information from exploitation.

4.3 MAKING AN ACTION PLAN

4.3.1 PREPARE THE MATERIALS

If the community wants to make a scale map, then decide how to obtain the necessary tools and learn the skills required. With determination and creativity, it is always possible to obtain the tools that you need. Maybe the community wants to start a fund in order to buy compasses. Maybe the community is already working with a non-governmental organization (NGO) who can help to find funding to buy tools. Or maybe it is possible to borrow tools from an NGO or from a neighbouring community. If it is not immediately possible to get the materials to make scale maps, then start by first making sketch maps.

Use the table below to help make a list of the materials that you will need to make the kinds of maps that you've planned. Then discuss with the team or the community how to obtain the materials.

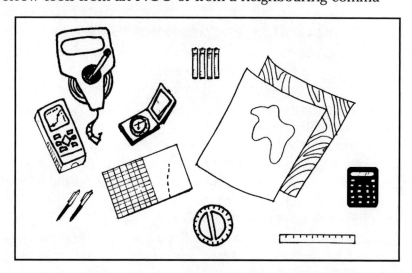

MATERIALS NEEDED FOR DIFFERENT MAPPING TECHNIQUES

GROUND SKETCH	Any local materials: leaves, sticks, rocks, shells, hat, rope, cooking pot
SKETCH MAP	Pencil, eraser, pen, colour pens, large paper
SKETCH ON TOPOGRAPHIC MAP	Topographic map, pencil, eraser, colour pens, transparent paper, correction fluid
THREE-D MODEL	Topographic map, carbon paper, carton, plywood, colour pens, scissors, plaster, paint
FIELD SURVEY WITH GPS	GPS receiver, notebook or field forms, pencil, eraser, topographic map or graph paper, ruler, calculator, correction fluid
FIELD SURVEY WITH COMPASS	Compass, metre tape, clinometer, notebook or field form, pencil, eraser, calculator, graph paper, ruler, protractor
REMOTE SENSING	Air photos or radar or satellite image, stereoscope, special pencils, transparent sheet, ruler

4.3.2 ORGANIZE TRAINING FOR MAP-MAKING

Making scale maps requires some basic training. In order for a participatory approach to mapping to work, the villagers will need the skills to participate. Organize a small training session in the village, with the objective that villagers will be familiar enough with surveying and drawing maps to participate fully in the mapping activities.

Such a training session will take two or three days. Afterward, motivated individuals will continue to learn both theory and practice as they carry out the mapping activities.

An outside trainer can be invited to the community, or a few community members can be sent for mapping training elsewhere and return to conduct the training in the village. Those people who are sent for training should be carefully selected for their capacity to learn, for their ability to teach others, and for their commitment to use their knowledge to facilitate a participatory mapping project to completion. It is best to select people who are motivated, have a high-school education, are eager to learn new things, and are interested in community land issues.

4.3.3 SCHEDULE THE MAPPING PROJECT

It's often best to think of mapping as an ongoing project. For example, some indigenous groups in Canada took 15 years to make maps of their traditional territory and they are still continuing to add more information to their maps. Stories about the land have taken generations to be created, told and retold. Therefore, mapping those stories accurately will also take time.

Before starting to map, though, it's important that people in the community know approximately how long the basic mapping will take, and for what activities, or stages, in the project their involvement is needed. Then the community as a whole knows what their investment is in the project, people can agree on a schedule, and individuals can schedule time to participate.

Scheduling is made more difficult because the number of days or weeks to complete a particular part of the mapping can vary greatly, depending on

❖ *The size of the area*
❖ *The mapping technique(s) used*
❖ *How much detailed thematic information is collected*

MAPPING SCHEDULE: AN EXAMPLE FROM INDONESIA

This example will help you to estimate the time needed: Consider a typical community mapping project in Indonesia by and for a community of 400 people, living in an area of 10 km × 10 km, drawing 3-5 different thematic maps. Their schedule looks like this:

ACTIVITY	TIME REQUIRED (days)
Preparations and community meetings	2–3
Training in the village	2–3
Field surveying by day and gathering local knowledge in the evening	6–15
Processing the data and producing the final map	6–12
Presentation of the final map and discussion of follow-up activities	1
Total	**17–34**

STAGES IN A COMMUNITY MAPPING PROJECT

INTRODUCTION TO THE IDEA OF MAPPING
Informal discussions in the community

IMPORTANT COMMUNITY DECISIONS
First community meeting
☞ What are the purposes for mapping?
☞ Whether or not to make scale maps?
☞ Who will control the finished maps?
☞ Who will make up the community mapping team?

Output *(optional)*
☞ Letter of commitment by village council
☞ Letter of request to an assisting organization

PLANNING THE MAPPING ACTIVITIES
Second meeting of the community or the mapping team
☞ What information to map?
☞ What area to map and at what scale?
☞ Who will do the training?
☞ How to organize community participation?
☞ When to do the mapping activities?

Output
☞ Sketch map, for the purpose of documentation or planning to make scale maps
☞ List of information to map

FIND AND PREPARE A BASE MAP AND TOOLS
Preparations in the city

TRAINING FOR COMMUNITY MEMBERS
Presentations in a house or school and practice in the field

Output
☞ Villagers have skills, knowledge & confidence

PARTICIPATORY MAPPING OF LOCAL KNOWLEDGE
Surveys in the field and gatherings in the village
Any combination of techniques:
☞ Make 3-D model and discuss it.
☞ Sketch directly on topographic maps.
☞ Survey with GPS (boundary and important places).
☞ Compass survey of village area and roads and trails.
☞ Use participatory activities to draw local knowledge on maps.

Output *(optional)*
☞ Quick scale map (on topographic map)
☞ 3-D Model for ongoing discussions

PRODUCE THE FINAL THEMATIC MAPS
Probably in the city
☞ Process survey data.
☞ Compile survey maps & sketch maps.
☞ Design and draft the final maps.

CHECK & VALIDATE THE MAPS
In a community meeting
☞ Revise the maps as necessary.

Output
☞ A set of thematic maps of local knowledge, made and validated by the community

Schedule the mapping activities for when most villagers have some free time, perhaps after the harvest or after planting, or when the fishing season is closed. If you have time, make a seasonal calendar of community activities (see section 11.3). The calendar will help to schedule the mapping activities by showing when the least busy month is, and it will also provide useful data for the project.

5 MAKING SKETCH MAPS

What you will learn in this section:
> *What sketch maps are and what they are for*
> *How to make a sketch map on paper*
> *How to make a ground map for discussions in the village*
> *How to make transect sketches by drawing the route you walk*
> *How to make a panorama sketch from a viewpoint*

5.1 SKETCH MAPS

5.1.1 WHAT IS A SKETCH MAP?

A sketch map is a drawing of the land from a bird's-eye view, or as if looking down from an airplane. Sketch maps are pictures of local knowledge, made by hand and from memory, usually drawn on paper with pencil or coloured pencils. To make a sketch map does not require any measurements, calculations, special tools or technical knowledge. Sketch maps are not made to scale. Thus, as a final product, sketch maps are not very useful when quantitative accuracy is important—when you want to determine the size of an area or make other quantitative analyses. And, because of this limit, sketch maps may lack credibility with government officials.

However, local villagers have more in-depth knowledge about the local area than do outsiders, though the latter may have sophisticated survey tools. A sketch map drawn by villagers from their own knowledge can be very detailed—and it might be a truer reflection of what the land means to local people—than what outsiders might depict on a scale map. In this sense, from the local people's perspective, a sketch map is a more accurate picture of their land than a technical map. There is a tendency to believe that the more sophisticated the instru-

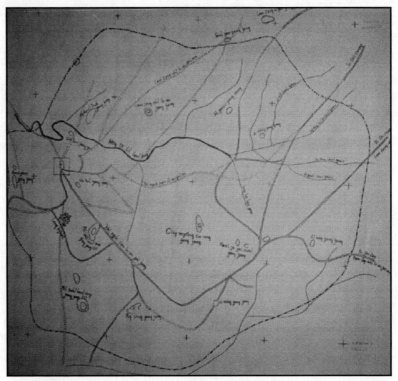

Photo of a sketch map.

ments we use in making a map, the more accurate it is and the better it is, but whether that premise is true or not depends on the purpose of the map.

5.1.2   WHAT ARE SKETCH MAPS FOR?

Sketch maps are excellent tools for internal community discussions about land-use conflicts, environmental problems, and land-use planning. In fact, a sketch map is usually more effective than a technical map for internal discussion within the community because sketch maps are easy for everyone to understand, even without training.

Sketch maps may also be sufficiently useful for initial discussions with outsiders, for instance about land-use conflicts. Sometimes just a clear picture, not necessarily a quantitatively measurable one, is enough to clarify the issues and allow disputing parties to better understand each other's point of view, so that an initial agreement can be reached.

Sketch maps can be important historical records. In many places, decades ago, foreign explorers, missionaries, anthropologists, and indigenous peoples themselves drew sketches of the peoples' customary lands. In Malaysia, Canada, and Australia these maps have been used as evidence in court cases regarding land rights.

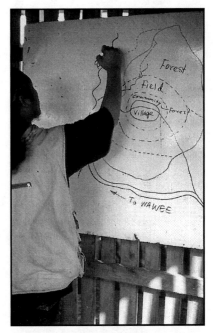
Draw a sketch map to help plan the technical mapping.

Making sketch maps is also an essential early step in community-based mapping. Sketch maps are useful in a technical mapping project to

❖ *Record the villagers' perspective of the land*
❖ *Enhance interviews and communications about the land*
❖ *Become familiar with the landscape (if you are an outsider)*
❖ *Plan the survey*
❖ *Quickly record information in the field while you are doing a survey*

A sketch is an important record, regardless of whether you draw it for the purpose of creating a temporary discussion map, as a final map, or to take notes while surveying. Therefore, write a date and title on it and store it safely.

5.1.3   MAKING SKETCH MAPS

What we commonly refer to as a sketch map is a picture of the land drawn, as if from above, using pencil on paper. In community-based mapping, sketch maps are often drawn from memory by groups of people while sitting in the house.

Often, the most difficult step in drawing a sketch map is facing the blank paper and drawing the first line. Try starting from the village centre and drawing outward from there. Or start by drawing the

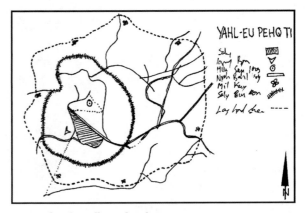
Example of a village sketch map.

main river that crosses the territory. Then draw in the tributaries one by one and name them. Finally, draw and name the mountain where each tributary emerges.

Whether you are making a sketch map as a tentative map or as a final map, it is most important to include many natural landmarks, because landmarks help people to locate the map on the ground or to find the area on a topographic map.

Name all the features and places on the map—such as rivers, ridges, mountains, valleys, rapids, trails, farming areas, etc. Use specific colours or symbols for kinds of features that occur frequently on the map, whether they are areas or points. Make a legend on the side to explain the meanings of the symbols.

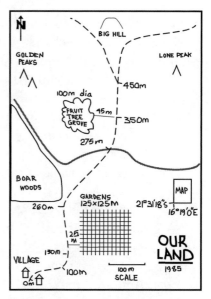

An example of a sketch map made by pacing to estimate distances.

Although sketch maps are usually drawn with only relative direction and distances, you can achieve slightly more accuracy by estimating distance and the direction of each feature in relation to others. Your body becomes an important tool. Distances on the map are estimated using the length or width of your fingers, or the span of your hand. Knowing the equivalents in centimetres for several of these units allows you to measure in the field when no other devices are available. Pacing is one method for measuring short distances on the ground (see the sidebar 'Estimating Distances' in subsection 9.3.2). Longer distances may be estimated by timing how long it takes to walk the distance, using a watch or some other indicator (such as the time it takes to sing a certain song), or to smoke a cigarette (less accurate).

Another way to make a sketch map a little more accurate is to use a small-scale topographic map. It might show only the main river through the territory, but use it as a framework on which to sketch details. (Consider enlarging the relevant part first.) As long as you can accurately locate the village on it, this kind of map at least shows where the territory is located within the province, state, or country.

In participatory mapping we want to encourage input from all sectors in the community. Each sector has different mental maps. Hunters probably travel to the outer boundary of the territory and know more about animal habits, whereas women often stay closer to the village and know more about the farm fields and medicinal plants. To draw sketch maps, the villagers may divide into small groups of women, youth, elders, hunters, etc. People often find it easier to talk and participate in small groups than in large groups. Moreover, because in some cultures women

Young people sketching a land-use map.

are not accustomed to expressing themselves in groups that include men, and because youth often defer to elders, in a large group the women's and youths' knowledge and perspectives might never get sketched.

A facilitator can help the process by suggesting where to start drawing (for instance, a major river) and by asking questions of the villagers who are drawing the map. Questions such as the following can help to jog the memories of the map-drawers:

❖ *Where is the mouth of that river?*
❖ *How far upriver is the next tributary?*
❖ *What is the name of the mountain at the headwaters of that river?*

5.2   MAKING A GROUND MAP

A ground map is a map created outside on the ground using leaves, rocks, beans, wood, reeds, or any natural material available. It could also be made on the floor of a large meeting room using hats, shoes, rope, pieces of paper cut to shape, or anything you can find. A ground map is usually drawn very large, perhaps 10 m × 10 m. At such a size, many people can gather around, and all can see it and discuss it. Making a ground map is a highly participatory activity and can be like a game. People can walk on the map, helping them remember the details about the route to their garden, etc.

To make a ground map, first find a large, open space. Villagers can become accustomed to mapping by first mapping the village rather than the whole territory. Place a rock or other marker to represent a central land-mark, such as the church. Then a facilitator could ask the villagers to draw or otherwise mark other important landmarks.

Obviously, such a large representation cannot be a permanent record, thus ground maps are some-times called *ephemeral maps* or *temporary maps*.

Making and discussing a ground map.

When the map has been completed to the satisfaction of the group, one person can draw it on a piece of paper or in a notebook to keep a record of the map. Also record details such as the names of the house owners.

If you've still got enough space on the ground, continue drawing the rest of the whole territory. Otherwise, erase the village map and start a new one of the territory. This time use a rock (or other object) to represent the centre of the village. Again, have someone draw the map on paper once it has been finalized.

5.3 OTHER TYPES OF SKETCHES

5.3.1 TRANSECT SKETCH—A BIRD'S EYE PERSPECTIVE

Like a sketch map, a transect sketch is drawn from a bird's-eye perspective. The difference is that you draw only what you can observe on either side of you while travelling along a route. On paper, the sketch will always look long and thin. You could make a transect sketch while walking along a road or along a trail through gardens, or as you travel down a river in a canoe.

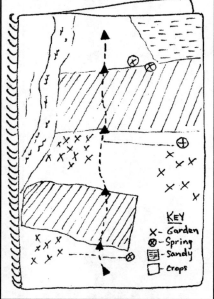

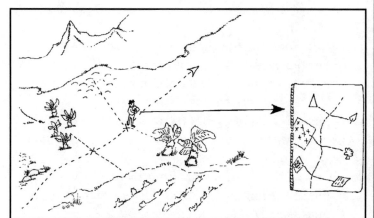

A transect sketch is made by drawing what you see on both sides of you as you walk along a road or trail.

A transect route is often chosen to cut across the land in order to get an idea of the diversity of land-use types. Choose the route according to the purpose. If the purpose is to see where different land-use types are located and the area is mountainous, then a transect will probably show the greatest diversity if the route goes from the lowest to highest place in elevation. If the purpose is to show distribution of land ownership, then you would choose a transect route that passes the greatest variation in ownership.

Draw transects starting from the bottom of the page. Include natural features that might allow you to locate the transect on a topographic map. If you draw the sketch on graph paper, you can choose a scale, for example 1 cm represents 20 m. Estimate distances on the ground and estimate distance on the sketch according to your chosen scale. If necessary, use several sheets of paper and combine the information later.

5.3.2 TRANSECT SKETCH—A PROFILE PERSPECTIVE

The same transect route for which you might draw a bird's eye transect sketch can also be drawn from a profile or horizontal perspective. This kind of profile can clearly show what kind of land use takes place near the river and on the mountain. Or what kind of land use takes place close to the village and farther away. Different zones become evident. Leave space below the profile drawing to write notes about the different zones. (See illustrations on next page.)

5.3.3 PANORAMA SKETCH

A panoramic sketch is a picture of the landscape as seen by the human eye—usually from a viewpoint, a place where you can see a fair distance. Panoramic sketches are

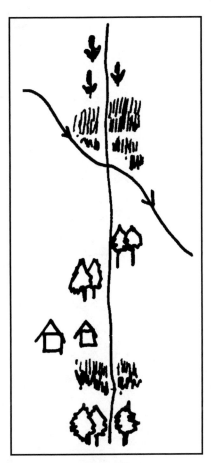

usually drawn about twice as wide as tall. Use a left-hand page of your notebook for drawing and the corresponding right-hand page for notes. Use pencil so that corrections are easily made (ink it in later, if you wish). There are seven steps in drawing a panoramic or landscape sketch:

1. **Select a view** that shows contrast in topography or land use, preferably with features that can be identified on a topographic map.
2. **Turn the book sideways and draw a frame** to help you to fix the vertical and horizontal extent of the picture.
3. **Mark your position** at the bottom centre of the page. Give your position a station number (if it does not already have one) so that you can record it on your field map and reference it in your notes.
4. **If possible, identify the location of the viewpoint** with reference to the base map, or with triangulation bearings to known peaks.
5. **Draw in the outlines of the hills** as you see them. Estimate the angle of vertical (use something straight, such as the edge of the notebook).

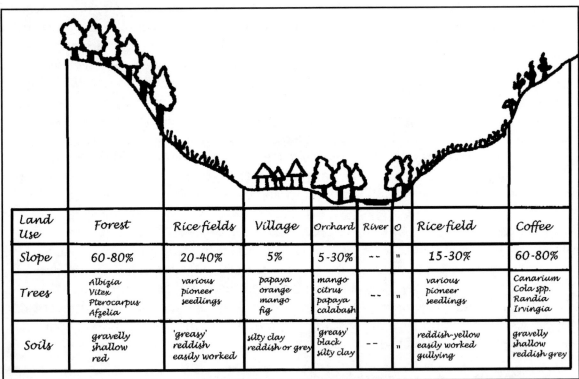

Land Use	Forest	Rice fields	Village	Orchard	River	o	Rice field	Coffee
Slope	60-80%	20-40%	5%	5-30%	--	"	15-30%	60-80%
Trees	Albizia Vitex Pterocarpus Afzelia	various pioneer seedlings	papaya orange mango fig	mango citrus papaya calabash	--	"	various pioneer seedlings	Canarium Cola spp. Randia Irvingia
Soils	gravelly shallow red	'greasy' reddish easily worked	silty clay reddish or grey	'greasy' black silty clay	--	"	reddish-yellow easily worked gullying	gravelly shallow reddish grey

A transect sketch (top), which is drawn from a bird's-eye perspective, shows what you see when you walk a given route. The information on a transect sketch (along with other information from your notes) can later be drawn as a profile perspective (bottom), which can be labelled to clearly show different activity zones.

6. **Draw and label bearings** *from the viewpoint to the major hills and river junctions using small arrows.*

7. **Draw anything else you choose,** *such as trees, rivers, roads, grass, or gardens. Put labels and notes on the sketch and on the right-hand page.*

Of course, you could also use photographs to record panoramas, but sketching has several advantages:

❖ *You can emphasize depth and slope perspective that may be lost in long-distance photographs.*

❖ *Sketching makes you observe closely.*

❖ *It's possible to eliminate obstructions in the foreground.*

❖ *You can select the most significant features.*

❖ *You can add notes to the sketch while the actual scene is in view, a more reliable method than annotating a photograph from memory or written notes after a lapse of time.*

❖ *Sketches are available for immediate reference at the end of the day.*

Drawing a panorama sketch while in the field.

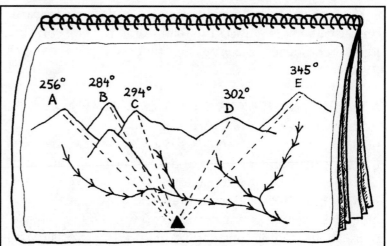

A panorama sketch without bearings and one with bearings.

PREPARING A BASE MAP

What you will learn in this section:
> ➢ *What a base map is used for*
> ➢ *That there are three ways to prepare a base map*
> ➢ *How to decide what scale to use for a base map*
> ➢ *How to change the scale of an existing map*
> ➢ *What to check when making a base map from a topographic map*
> ➢ *How to copy the base map and how many copies to make*

6.1 WHAT IS A BASE MAP?

A *base map* shows landmarks or reference features, such as rivers and mountains, that serve as a framework on which to draw *thematic maps*. You might find it useful

to think of a base map as being like the posts and beams of your house and the thematic information that you add to it (the locations of gardens or gravesites, for instance) as being like the thatch or boards and your furnishings.

The most common and useful type of base map is a topographic map. Topographic maps accurately show the shape of the land—valleys, mountains,

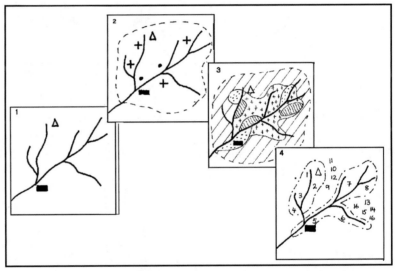

A base map (1) is the foundation for a set of thematic maps (2–4).

ridges, slopes, and so on—and the positions of bodies of water—such as rivers, lakes, and the ocean (and possibly other information). These landmarks are all reference features that someone using the map would see both on the ground and on the map. The location of locally known places, such as a sacred area, can be drawn in relation to these reference features.

6.1.1 THREE WAYS TO MAKE A BASE MAP

Use an already published scale map, such as a topographic map. Many kinds of published scale maps can be used as base maps. The most useful ones are topographic maps. Topographic maps accurately show the locations of lakes and rivers and the shape of the terrain. Using a topographic map gives you an inexpensive and easy-to-learn starting point for making accurate thematic maps—and you can map a large area quite quickly because it is not necessary to walk to every place on the map. (See chapter 7 for more about reading topographic maps.) Topographic maps are available for most

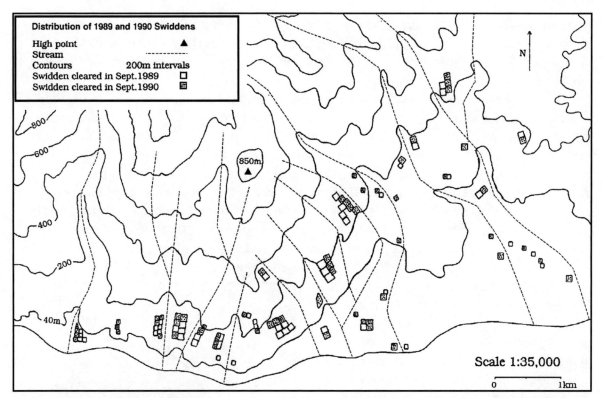

N

800

600

850m ▲

400

200

40m

Scale 1:35,000

0 —————— 1km

A topographic map makes an excellent base map on which to draw boundaries, land uses, or cultural areas—or on which to show local place names.

everywhere in the world, but in some regions it may be difficult to obtain maps of a scale large enough for community mapping. **Land survey maps** are scale maps that are produced for specific areas, for instance for mineral exploration. They show rivers, roads and ridges and are usually at a large scale. **Property title maps** might be used for a base map, although they show boundaries only.

Use remote sensing: air (aerial) photographs, radar images, or satellite images. Remote sensing information consists of pictures of the ground taken by sensors mounted on airplanes or satellites. Three commonly encountered forms of remote sensing are air (aerial) photos, radar images and satellite images. (Each of these kinds of pictures is described more fully in section 6.3.) Remote sensing images are very useful because they often show more information than a topographic map: they show thematic information, such as vegetation cover or land use, as well as the basic information, such as rivers and mountains. But, in order to make use of the extra information, you require some special knowledge and experience in interpreting the images. Also, remote sensing images usually have some distortion in the scale and need to be *georeferenced* (mathematically corrected) before

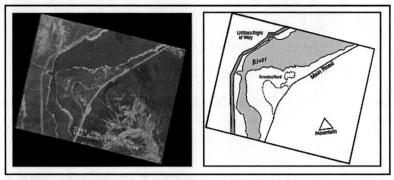

A base map, or even a thematic map, can be prepared from an air (aerial) photograph.

we can use them to make an accurate scale map. Most remote sensing products are more expensive than topographic maps, and in many areas they are not available. See appendix G for information on how to find topographic maps and remote sensing products.

Make a base map by field surveying with a compass and metre tape or the Global Positioning System (GPS). If you don't have access to a topographic map or remote sensing, then you don't have any choice but to make your own base map. You can do so by using a surveying tool, such as a compass or a Global Positioning System (GPS) receiver, to identify locations on the ground and draw them to scale on a piece of paper. Making a detailed base map with a field survey means that you have to walk around and across your whole territory. However, this method is very time consuming and the result won't be as detailed as a topographic map. Therefore, this approach is recommended only if you can't get a topographic map, and if the budget allows for sufficient surveying hours. For a large territory, it is probably not practical.

Regardless of which tools you use for surveying in order to make a base map (on which you can add details later), it is important to include many natural features. These features are the landmarks or reference points that are necessary to help you or anyone else locate themselves on the map in the future.

Of course, for efficiency—to save you having to traverse the same area again—while you are in the field you will record thematic information about the area as well. When you process the data to draw on the map, you would select the most basic information, such as the locations of rivers and roads, to create a base map. Other information (for example, place names and the locations of gardens) you would draw on separate thematic maps.

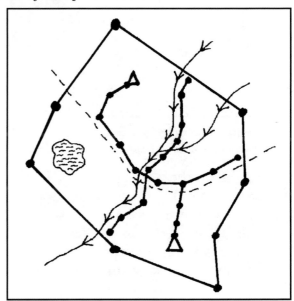

For a large area of land, surveying a base map is possible, but will be very time-consuming.

(Look at chapter 9 to learn about techniques for traversing with compass and metre tape, and see chapter 10 for how to use the GPS).

6.2   PREPARING A TOPOGRAPHIC BASE MAP

This section explains how to prepare a base map by using a topographic map—or from any other kind of published scale map.

6.2.1   OBTAIN A TOPOGRAPHIC MAP FOR THE AREA

Topographic maps are made at various different scales. Some standard scales are 1:250,000, 1:100,000, 1:50,000, and 1:20,000. Look for a topographic map with a scale as close as possible to the scale of the final map that you want to make. Remember that the smaller the map scale (the bigger the number on the right), the less detail. Community mappers of northern Canada and South America often use a small-scale map of 1:250,000 to show their territory. But that scale does not show enough detail for

Southeast Asian communities, who can show their whole land-base on a 1:10,000 or 1:50,000 scale map. Note that the largest scale topographic map available for many parts of Southeast Asia is 1:50,000.

Topographic maps are indexed. Contact the government's Lands and Surveys Department (or equivalent) and first obtain an index map for the map series at the scale you want. An index map is a small-scale map, perhaps for a whole province or state, that shows latitude and longitude and the major rivers. The

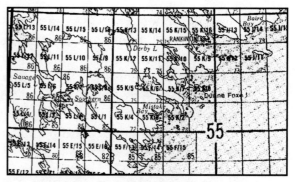

To order topographic map sheets, look on an index map to find the code or number for the map(s) that you need to cover the area.

squares with numbers inside them indicate the various map sheets available. On the index map, locate the area that you want to order maps for according to the major rivers that cross it or by taking a GPS coordinate on the ground. Order the map(s) by map-sheet number. (Maps for some countries have a name for each sheet as well—the name of a town, lake, or mountain within the map area, perhaps.)

You will frequently find that the area you are interested in is situated near the edge of the map, and it may even extend beyond the border of the map onto the adjoining map sheet(s), so be sure to order all the maps that you need for your mapping project. Once you get them, make photocopies of the originals. Cut away the margins and unneeded areas on the photocopies (don't cut right to the territory boundary—just cut the pieces rectangular). Then tape the sheets together to make one map sheet.

6.2.2   SELECT THE SCALE TO WORK WITH

It is always best to choose the scale with the participation of the community, in consultation with local people who know the area that they want to map. After making a sketch map of the community's land boundary, you can estimate the size of the territory and from that information choose a scale that is appropriate (see also subsection 4.2.2). If you already have a topographic map available, then search for boundary landmarks on the map. That way your estimate of the size of area will be more accurate than if you had used a sketch map.

The scale to use for a map depends on the area of land that is to be mapped and the size of the paper that it is desired to fit it onto. Be sure to take into account the space needed for the map's *margins* when you make this calculation. For example, if the land area is 5 km across and the paper is 100 cm wide, allowing a 10 cm margin (both sides) gives 80 cm of usable width into which you must reduce 500,000 cm.

Use this equation to calculate the scale that you need to work with:

scale = width of area (cm) / width of paper (cm)

For the example given above, scale = 500,000 cm / 80 cm = 6250, or 1:6250

Round the number off to a convenient *smaller* scale to ensure that you can get everything on the map. In this case you might choose to use a scale of 1:7000 or 1:10,000.

It is generally simplest to use the same scale for all the thematic maps for the village. However, you could choose to make maps at different scales without too much complication—it might be easier than splitting a map into several sheets. For example, all the agriculture and important cultural sites could be shown on a large-scale map at 1:10,000. You could make the other maps or overlays at this scale too, but if the local people travel great distances to hunt, it may be more convenient to use a 1:50,000 scale map to show hunting ranges. And perhaps the people's known historical migration is over a distance of 100 km. You'd have to show that information on a 1:100,000 or smaller-scale map.

6.2.3   CHANGE THE SCALE OF THE TOPOGRAPHIC MAP

Chances are that the scale of the government topographic map for the area is not at the scale that you want to work at. You have already decided what scale you want. Now that you've found the topographic map closest to the scale that you need, how do you change its scale to exactly what you need?

In most cases you want to enlarge the map; rarely would you reduce it. Here are three methods for enlarging maps:

❖ *By hand: Draw a small grid (say, 0.5 cm) on the original map and a proportionally larger grid (that would be 2.5 cm for a 500% enlargement) on blank paper, then use the grid to guide the drawing of the map features, square by square.*

❖ **Computer:** *If the map is already in the computer, a mapping software program can enlarge or reduce the map to any scale. Or you may choose to scan it with a scanning device and use a graphics program to scale it.*

❖ **Photocopy:** *There are large-size photocopying machines with the capacity for enlarging or reducing maps.*

The most common way to enlarge maps is with a photocopying machine, so the rest of the steps that follow will assume this method.

6.2.4   ENLARGE THE MAP WITH A PHOTOCOPYING MACHINE

The easiest way to enlarge the map is by photocopier. A photocopying machine enlarges by percentages. If you enlarge at 200%, you've doubled the dimensions of the map, or halved the scale, say from 1:50,000 to 1:25,000. So, if you want to make a 1:50,000 map into a 1:10,000 map, you need to enlarge the map by 5 times or 500%. If you want to enlarge a 1:100,000 scale map to 1:10,000, you must enlarge it by 10 times or 1000%. Remember the equation like this:

> **percent enlargement = (scale of the original / scale wanted) × 100%**

However, most photocopiers can't enlarge by 500% in one pass. They usually have a maximum enlargement of 150% or 200% (simpler machines may only enlarge at certain standard percentages—you may need to be creative—but sometimes the controls for the 'zoom' function are just hard to find).

Here is one way to enlarge a 1:50,000 map to 1:10,000 using a 200% machine:

Step 1:	1:50,000 × 200% → 1:25,000
Step 2:	1:25,000 × 200% → 1:12,500
Step 3:	1:12,500 × 125% → 1:10,000

Be sure to set the 'density' or 'light–dark' control so that you get both clear white areas and solid black lines, if possible. (The 'automatic' setting may not give good results, especially if you need to copy in several passes.)

When you enlarge a map on a photocopier, the paper size may be too small for the complete map. In this case you will need to work with a piece at a time. Glue or tape all the pieces carefully together when you are done. Be sure to check for improper overlap and skewed angles and make sure that all the pieces are flat, not twisted or bent.

6.2.5   CHECK THE DISTORTION OF THE PHOTOCOPY

Next, check the result of the photocopy process for whether
- ❖ *You calculated the percent enlargement correctly*
- ❖ *Any distortion was introduced by the photocopy machine*

To check the enlargement, use a ruler to measure between the same two points on the old and new maps. UTM grid lines (see subsection 2.2.4) are good for that because they are metric. If the UTM lines are 1 cm apart at 1:100,000, then they should be 10 cm apart at 1:10,000. Unfortunately, older or poorly maintained machines may enlarge unevenly—say 205% in one direction and 195% in the other, or 200% in the centre and 205% at the edges, so it's a good idea to check the map for distortion by measuring in several places, for example, at the centre and near each of the four corners. The greatest distortion is usually at the edges. Of course, every time you copy a copy, the distortion gets worse, so use the minimum number of enlarging steps possible.

Measure between grid lines both horizontally and vertically in several places on the map and calculate the distortion (use the equation below). If, at any place on the map, the distortion is greater than 3%, then the distortion of the enlargement is too great. Find a way to improve the process.

Calculate the percent distortion (D) for each measurement using this equation:

$$D = ((L_c - L_p) / L_p) \times 100\%$$

where L_c = measured length on copy, and L_p = enlargement factor × length on original

Example: A map was enlarged by a factor of 5 from 1:50,000 to 1:10,000.
L_c is measured to be 5.1 cm.
$L_p = 5 \times 1$ cm = 5 cm, so D = (5.1 cm – 5 cm) / 5 cm × 100% = 2%

Compare the calculated error for the width direction with that for the length direction and for other places on the map. If the error is more or less uniform, you need only adjust the enlargement setting on the photocopier. If the error is not uniform, the distortion will be too great—perhaps you can get adequate results if you cut off the edges at each enlarging step and only use the centre piece. If the width and length errors are consistent across the map, maybe rotating the paper by 90° at each enlargement will even things out. Or look for another machine.

6.2.6   CLEAN THE ENLARGEMENT

When a topographic map is enlarged, it creates holes in the lines or marks where you don't want them, and sometimes there is additional information on the map that you don't need. Take typewriter correction fluid (Wite-out, Liquid Paper, etc.—the kind sold for use with copies is best) and cover over the lines and marks that you don't need. Use a black pen to fill in the lines that have been broken or lost. If you have a really bad or cluttered copy, it might be simpler to trace just the information that you need onto a sheet of tracing paper and photocopy this information onto regular paper.

When enlarging a topographic map, you may have had to cut off all the original information in the margin, such as the legend, scale, map-sheet number, etc., or maybe it just didn't get copied. Create a new margin when making copies of the enlarged base map. Copy the original legend or make a new one. Mark the latitude and longitude and the map-sheet number on the new map. Make a new graph scale too, and draw the north arrow in proper alignment with the grid lines. Before you make copies of the cleaned-up map, check the scale of the map, as described above. Also confirm that no important features vanished during the enlarging process.

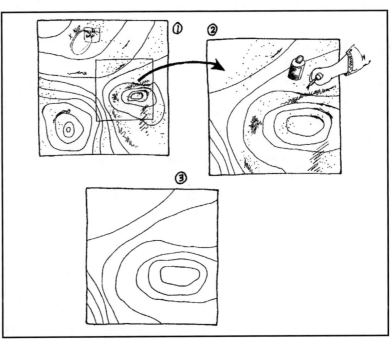

After enlarging a topographic map with a photocopy machine, hide unnecessary marks and lines using correction fluid.

6.3   USING REMOTE SENSING TO MAKE A BASE MAP

6.3.1   AIR (AERIAL) PHOTOGRAPHS

An air (aerial) photo displays exactly what your eyes would see from a window of an airplane: the shape of the land, the location of mountains and rivers, as well as different land uses and vegetation types. Most air photos are black-and-white, but colour photos are also available for some areas.

Air photographs are taken from an airplane using a specially mounted camera. Photographs are taken continuously as the airplane flies a straight course along each of a number of sequential parallel lines in turn. Because the airplane follows a known bearing, the orientation and the coordinates of the photo are also known. Because the airplane flies at a known distance above the ground when it takes the photos, it is possible to calculate the scale of the photos. In general, the higher above the ground the plane flies, the smaller the scale of the photo, though the magnification of the lens on the camera is also a factor.

Air photos are taken so that the image on one photo overlaps with the image of the adjacent photo. If you look at a pair of overlapping photos with a special viewing device called a *stereoscope*, then you can see the photo image in three dimensions (3-D). The heights of trees and mountains and the steepness of the slopes all become apparent, and it is even possible to calculate the heights of trees and mountainss.

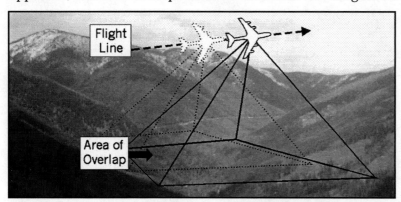

Air (aerial) photos are good for making base maps because they show landmarks (reference features) as well as land use and the condition of the land.

Whenever we are looking at air photos, we must remember that there is always considerable distortion of scale at the four edges of the image. This distortion is caused by the angle at which the photo was taken from the plane. At the edges it is impossible to accurately measure distances or establish the correct locations of features.

Survey departments sometimes sell air photos that have been *georeferenced*, or processed into air-photo maps. They have corrected the distortion in the photo and interpreted the topography with height calculations. Then they reprint the photo, showing the scale, orientation, coordinate system, and possibly contour lines on it. Because it is georeferenced and shows the photo image, an air photo map is an excellent base map for community mapping.

Unlike a map, an air photo has no legend to tell us what the objects are. We have to interpret what the objects are. The goal of air-photo interpretation is to differentiate objects on the photo; perhaps to identify the course of a river, the border of a patch of forest, or the border of a field. Be aware of the following five types of qualities as you interpret the features on an air photo (if necessary, you can field-check areas that still puzzle you):

❖ *Colour or tone:* On colour photos we can see that a farm field is a different colour than the forest, and a young forest is a different shade of green than an old forest. On black-and-white photos there are many tones of grey. An old forest may be darker grey than a young forest, for example.

❖ *Shape or form of objects:* Some objects, such as airstrips, have distinctive shapes.

❖ *Texture:* Some objects appear to have a smooth surface, whereas others look rough. A dense forest will appear smooth, but a widely spaced tree plantation will look rougher and show a pattern of lines.

❖ *Association/location:* Some objects can be identified by association with other objects. A smooth-textured area of vegetation at the edge of a lake and with no road or farming nearby we could assume is a swamp rather than a farm.

❖ *Pattern:* Some objects can be identified by the overall pattern. It is easy to distinguish between natural features and those created by people. Features created by people are usually linear and geometric, whereas natural features are seldom straight.

6.3.2 RADAR IMAGES

Radar images, like air photos, can also be taken from an airplane, and the result is a black-and-white image of the ground that looks similar to an air photo. But, unlike an air photo, radar images are taken with radar wave equipment rather than a camera. The difference is that a camera is a *passive sensor* that receives natural sunlight reflected by objects. Radar is an *active sensor* that operates by sending out a signal that is reflected back from the features of the landscape. Radar is not obstructed by clouds, so it is particularly useful in tropical or mountainous regions with persistent cloud cover, or where dense vegetation, smog, haze, or smoke obscure the ground.

Radar images are interpreted using the same principles as air photos, but may require some additional training to interpret. Not all land features are obvious from the image. This problem is partly because of deceptive shadows in areas that the radar missed because of the angle at which the sensor was set to take the image. Shadows caused by large objects such as mountains and tall buildings can completely obscure other features. Scan lines caused by the radar waves scanning the ground can also obscure some features in the image. Also, as with air photos, the scale is distorted at the edge of the image.

Radar images are taken systematically, like air photos, so they too can be processed into a map (a *radar map*) with correct scale, orientation and coordinates.

Radar images can also be taken from a satellite, as shown in this example.

6.3.3   SATELLITE IMAGES

Satellite images are made from data acquired with computers from sensors on satellites. The data is processed by computer to form an image. You can order a satellite image that has already been georeferenced and processed to look like a colour photo of the land. Alternatively, you can order the computer data (digital data), and—with computer equipment, software, and training—process the data yourself to make an image.

There are two well-known types of satellite imagery: LANDSAT and SPOT. Each LANDSAT image covers an area 185 km wide by 170 km long. The **resolution**, or the smallest object that you can see on the image, is 30 m × 30 m. This information means that a road or a field that is narrower than 30 m in width cannot be seen on the image. A SPOT image covers an area 117 km wide and has a resolution of 10 m × 10 m.

Because satellite images don't show as much detail as air photos, they are not very good for large-scale mapping (under about 1:20,000 scale). However, satellite images are excellent for small-scale mapping and especially for up-to-date information about the condition of the land. Satellite data is constantly being received at LAND-SAT and SPOT stations, so you can get satellite images for any day of the year. But the sensors can't get accurate data through cloud. Because the days are rare when there is no cloud over tropical areas, it may be hard to get useful images for them.

From http://users.powernet.co.uk/mkmarina

An example of a satellite image.

6.3.4 PREPARING BASE MAPS FROM AIR (AERIAL) PHOTOS OR RADAR IMAGES

Preparing a base map from air (aerial) photos or radar images is easy.

First, if the area that you want to map requires more than one photo, (and the photos have not been georeferenced), arrange the overlapping photos and tape them together so that you will be tracing only the central area of each photo, thereby avoiding the not-to-scale edges of the photos.

For the next step, if you have transparent plastic sheets (see section 12.3), you can lay one on top of the photo(s) and use erasable colour pens for the next step. Otherwise use special soft-erasable colour pencils for drawing directly on the photo(s). Begin by drawing the perimeter of the area that you want to map. Then highlight all the important reference features, such as rivers, roads, and mountain peaks.

If you drew on a plastic sheet, you can take it off and use a photocopy machine to enlarge the map to the scale that you want, as explained for topographic maps in sub-section 6.2.4. Otherwise, trace the features onto a tracing paper sheet and make the enlargement from it.

6.4 MAKING MULTIPLE THEMATIC MAPS FROM A BASE MAP

The main purpose of preparing a base map is so that we can use it to draw several thematic maps. A common approach is to make many photocopies of the base map, one for each kind of information or theme that you want to draw. Alternatively, use just one copy of the base map and draw each thematic map on a sheet of tracing paper or plastic. The objectives and skills of the mapping team will help you decide which way is best.

6.4.1 REPRODUCING THE BASE MAP

Note: before reproducing the base map, read subsection 6.4.3 to help you decide if you need to make any changes first.

Blueprinting is the least expensive way to reproduce large-format paper. First photocopy the clean base map onto special translucent blueprinting paper, and then make several blueprints. Photocopying the maps is easier and makes cleaner copies, but it is a little more expensive (reproduction methods are discussed again in subsection 12.2.1).

How many copies do you need? It depends on how many different thematic maps you want to make for this community, and how you plan to collect information. Usually, you will draw thematic information directly onto a copy of the base map. If, for example, you are making a map biography for each hunter in the village, then make a copy of the base map for each hunter. If you want to make three thematic maps—say, land use, cultural sites, and a boundary map—then use three copies of the base map and draw directly on them.

Make a few extra copies of the base map, in case you think of other information categories that you want to map, or in case the villagers work in small groups.

6.4.2 DRAWING THEMATIC MAPS ON TRACING PAPER

An alternative to drawing directly onto copies of the base map is to draw thematic information on tracing paper or clear acetate (plastic) sheets laid over the base map (see section 12.3 for more on different kinds of papers).

Using this method, you may need only one or two copies of the base map itself (think about how many maps will be worked on at one time) and lots of blank tracing paper or plastic sheets. Make registration marks with four precise small '+'s at the corners of the base map and on each tracing paper overlay, so that after you take the tracing paper off, you can line it up exactly again if you need to later. Find a flat working surface (such as a table, a board,

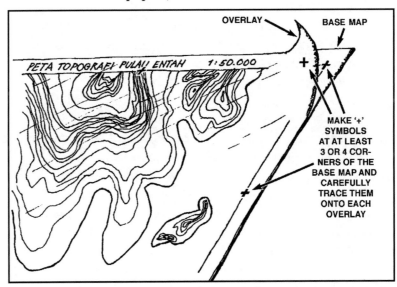

Overlays must be carefully 'registered' so that they line up properly over the base map.

or a sheet of hard plastic) so that you can align the base map and the tracing paper smoothly and exactly each time. Masking tape works well to hold the sheets in alignment. Thumbtacks (drawing pins) could work, but the holes tend to get bigger and bigger, making the registration less and less accurate.

The tracing-paper method is more awkward to use in the field or in the village than drawing directly on paper base maps. You can't fold tracing paper (roll it if you have to), and it distorts when it gets wet or damp (use a large plastic bag to protect it), and a flat working surface is more necessary.

6.4.3 MAKE THE BASE MAP INTO A LOCAL REFERENCE MAP

After a base map has been ground-checked and locally known landmarks and place names have been added, it becomes a local *reference map*. A reference map is a base

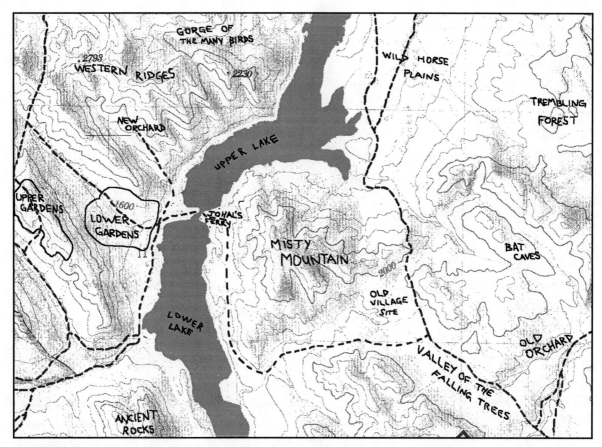

Adding local names to a topographic base map makes it more meaningful to the community.

map made more locally relevant. The more detailed and locally relevant the reference map, the easier it is for local people to draw thematic maps.

To ground-check the base map, take it out on the land and make sure that the features are accurately located and shaped. You will often find that some of the human-made features (such as villages, roads, and bridges) are missing or drawn in the wrong location. You may also find small rivers that are missing or have the bends in the wrong places, especially in heavily wooded areas or where there is a lot of cloud cover much of the time.

One way to quickly field-check the base map, if practical, is to climb to several viewpoints from which to take compass bearings to surrounding peaks and other features. By getting an overall view of the shape of the land and using compass triangulation (subsection 8.3.4) you can check whether the river headwaters are in correct relation to the peaks, and check the locations and angles of the smaller rivers and dry gullies. Modify the map if necessary.

If, according to local knowledge, there are many natural features missing—such as small rivers—then use basic field-survey methods to locate and draw them on the map. A quick survey can be done with a compass and paced distances.

A more rigorous way to check the map is to start the compass survey at a distinct starting point, such as a river junction. Then follow the rivers and roads with compass and metre tape to make an actual compass survey (see chapter 9). Record where small rivers enter and take bearings of river angles as they enter. This procedure, which is very time-consuming, is not necessary unless you know that the map has a lot of errors.

Check and correct the place names on the topographic map. Topographic maps are often made by government surveyors and cartographers who didn't ask local people for the names of places but instead gave places new names. Or the surveyors may have asked the people but did not hear or write the local names correctly. As a result, the names of places as shown on a topographic map are rarely the ancient names given to them by the ancestors of the community. Correct the names that are wrong and add names that are not written on it. The local names are important reference points for local people as they draw thematic maps.

Use the topographic map as a basis to draw important features of the land.

7 USING TOPOGRAPHIC MAPS

What you will learn in this section:
➤ *How to read topographic maps without using numbers*
➤ *How to read contour lines on topographic maps*
➤ *How to orient a map to the land by reading the features*
➤ *How to use a topographic map to make a three-dimensional model*
➤ *How to sketch thematic information on a topographic map*

7.1 ABOUT TOPOGRAPHIC MAPS

Topographic maps show the shape of the terrain with sets of curving lines. These lines are called *contour lines*. The pattern of the lines at any particular place tells you what the shape of the land is there. For example, you can see where the valleys are and where the ridges and the mountain peaks are, and you can see the arrangement of the rivers and the directions in which they flow.

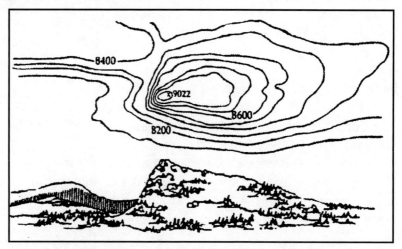

Topographic maps are characterized by contour lines that correspond to elevation and thus depict the shape of the land.

Topographic maps have a defined scale, a north direction, and a coordinate system (review coordinate systems in subsection 2.2.4). These items are identified in the map legend, along with the symbols for features such as administrative boundaries, settlements, roads, and airports. Because topographic maps are rich in basic information about the land and they are standardized, they make excellent base maps. Governments and industry use topographic maps as base maps for all kinds of mapping activities, such as land classifications and zoning, for resource inventories, and for research.

When you read topographic maps, it helps to understand how they are made.

REFERENCE

Railway, single track	
Railway Station, stop	
Main highway	
Secondary road	
Other roads	
Trail or track	
Boundary, international	+++++++
Boundary, county or district	
Boundary, park or reserve	
Icefield. glacier	
Marsh or swamp	
Tidal flats	
Rock...........................	+
Contours (in metres)	500
Bathymetric lines (in metres)	1000

The legend of a topographic map.

Topographic maps are drawn from aerial photographs. The exact location and altitude of the airplane is known when each photograph is taken. The photographs are then analyzed using special visual instruments that allow the photographs to be seen in three dimensions (3-D). The horizontal scale of the photo and elevations on the ground are calculated by knowing the distances and angles from the plane to the ground.

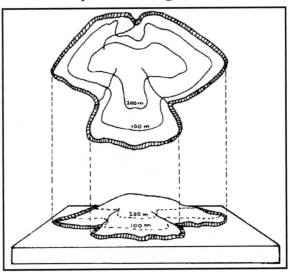

The locations of natural features are usually very accurate, but some of the information may be wrong or incomplete. Here's how it can happen: When you look at an aerial photograph, even if the land is covered in forest, you can usually see the shape of the land. You might see a small valley or gully in the topography and assume that there is a creek there, even though you can't see the water (but tree heights may be deceiving). But in flat areas it can be difficult to see the shallow valley created by a watercourse, so the map-maker might miss drawing some small creeks. Other things that are difficult to interpret on an aerial photograph, and therefore might be missing on a map, are caves or rock outcrops that might be covered by forest.

It takes practice to see the shape of the land in the contour lines of a topographic map.

Some features on the map might have changed since the map was made. Human-made features in particular—such as logging roads and buildings, or even small villages—may be newly developed or changed and thus not appear correctly on the map.

7.2   READING TOPOGRAPHIC MAPS

7.2.1   DIFFERENTIATING THE LINES

First, look at the map for an overview. Look at the map as a whole before dissecting it for analysis. At first glance, a topographic map looks like a mess of lines, especially if it is a black-and-white photocopy. If you have the colour original, keep it handy for reference, because it is much easier to read, and its coloured areas may also show features—such as open land and icefields—that may not appear on your black-and-white copy.

Second, identify the river network. The rivers are the most obvious features to find on the ground and on the map because they are represented by single lines rather than by the multiple lines that indicate ridges or mountains. River lines always cross contour lines, except where it is very flat and there aren't any contours; usually they do so where the lines indicate a valley or gully (explained next). Roads, on the other hand, which are also shown with lines, can cross contour lines too, but generally do so at a much smaller angle, and roads are often straighter than rivers.

Third, if the map is a black-and-white photocopy, use colour pencils to help differentiate the lines. To make it easier to see the difference between the river network and the contour lines, colour the rivers using a blue pencil. Start by colouring the largest river first and then follow its tributaries. Follow each branch back upstream from the largest river junction, then do the same for each of the small rivers feeding in. Colour roads red. Some topographic maps have lines around different vegetation types. Colour these lines green.

7.2.2 READING CONTOUR LINES

Contour lines connect points that have the same elevation. They show the height of the ground above sea level, as measured in metres or feet. Contour lines are drawn at a fixed interval—for example, at every 50 m of elevation, or at every 500 m. The contour interval is selected by the cartographer based on the map's scale and the ruggedness of the land. A 1:50,000 scale map usually has a contour interval of 25 m (20 m, 100 feet, and 250 feet are also common). A 1:100,000 scale map typically has an interval of 50 m. The contour interval is written in the map legend.

Usually, every fourth or fifth line, depending on the contour interval, is drawn thicker. These lines are called *index contours*. For example, if the interval is 25 m, then the lines at 100 m, 200 m, and 300 m would be shown thicker. Often, the elevation is written along these lines, but sometimes others are chosen. The elevation of a mountain peak is written exactly, and its location is usually marked with a dot.

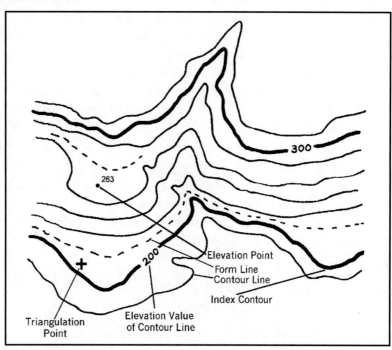

Learn to interpret the lines and symbols on a contour map.

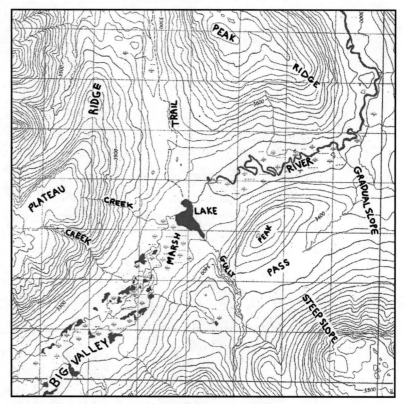

The pattern of the contour lines describes the landform.

Quickly interpreting topographic maps by the pattern of lines is easy if you remember these points:

- ❖ **Contour lines never cross each other.**
- ❖ **Rivers always cross contour lines,** *if there are any, because water flows downhill.*
- ❖ **The top of a mountain looks like a small circle** *or distorted circle.*
- ❖ **Widely spaced lines indicate a gentle slope and lines that run close together (and possibly disappear briefly) represent a cliff.** If you think of contour lines as being like stairs, a gentle hillside on the contour map looks like a staircase with deep treads (for people with long feet) when viewed from above. A very steep hillside looks like a staircase with very narrow treads (for people with short feet). A ridge has the appearance of two staircases back to back.
- ❖ **A valley or a gully makes a pattern of stacked 'V's or 'U's,** *depending on the sharpness of its bottom.* The point of the 'V' or curve of the 'U' points in the uphill direction. If the map shows a small gully, even if there is no watercourse shown on the map, you can predict that there may be one in real life, though it could be seasonal rather than year-round.

When you become familiar with reading contour lines, you begin to see how the patterns model the topography or shape of the land.

PRACTICE READING CONTOUR LINES

Take any topographic map and try to find these features:

☞ *Major river*   ☞ *Small gully*

☞ *Steep slope*   ☞ *Ridge*

☞ *Mountain peak*   ☞ *Small river*

☞ *Large valley*

7.2.3   ORIENTING THE MAP TO THE LANDSCAPE

A map is *oriented* when the features on the map are lined up in the same direction as the features on the ground. In the field it is easier to use the map if you first orient it. If you know where you are on the ground and on the map, simply turn the map so that the visible features on each correspond visually. Some features are better for orientation than others, because they are more permanent, more prominent, or more

Orient the map and yourself by aligning landmarks on the map with those on the ground.

distinct. A pointed hill is better than a patch of forest, for instance. (Another way to orient a map, with the help of a compass, is covered in subsection 8.3.1.)

Using just the map with its contour lines, you can often make a good estimate of where you are. Look around you and find the major ridges and mountains. Look for the tallest mountains. Look for river sources and the relative sizes of the rivers. Also look for the number of branches of each river. If you are near a river, which branch are you standing near? Locating yourself on the map is a matter of putting all this information together, of matching up the big pattern on the land with the little one on the map. Doing it easily can take practice—often it helps to move to a location with a better view. Of course, it is easiest if you are in an area with easily seen, distinctive sharp peaks and ridges and much more difficult in flat areas where the contours are not as obvious.

7.3   MAKING A THREE-DIMENSIONAL MODEL

A topographic map is flat, or two-dimensional, but it can be used to build a proportionally correct three-dimensional (3-D) model of the land. A 3-D model is easier for most people to understand than a topographic map, because it gives us a more realistic picture of the shape of the mountains and valleys. A 3-D model can be used for interviewing elders about the land, or it can be used to help a mapping team to better understand the topographic map.

Making a 3-D model is easy if you follow these seven steps:

1. *Determine the first contour line. Study the map carefully. The first contour line is the one that has the lowest elevation and is usually the line closest to the most downstream part of the biggest river (or the ocean) on the map. Trace this contour line with a coloured pencil. Note: The first contour line will usually not go all around the area of the map—it will appear only in one or several places on the map and will not form a closed path before it runs into the edge of the map. In this case, use the edge of the map on the **uphill** side of the contour line to complete the path, being sure to include in the path all the pieces of the contour line that go to the edge of the map. (And do the same for any higher contour lines that also go off the map.) If you find that the first contour line appears in several parts of the map as separate closed paths, deal with each enclosed area separately in steps 3 and 4. It is also possible that the land that you are mapping has a depression or lake without an outflow at its lowest point. In this case, when you get to step 3 you will need to begin with a piece of cardboard the size of the plywood, and the first contour line will indicate where you need to cut a hole in it.*

MATERIALS NEEDED TO MAKE A THREE-DIMENSIONAL MODEL

☞ *Topographic map (enlarged to an appropriate scale—to comfortably fill the plywood—if necessary)*

☞ *Sheet of plywood of manageable size (about 1 m × 1 m)*

☞ *Sheets of cardboard carton all of the same thickness*

☞ *Glue*

☞ *Scissors*

☞ *Small knife*

☞ *Carbon paper*

☞ *Plaster*

☞ *Paint and brushes*

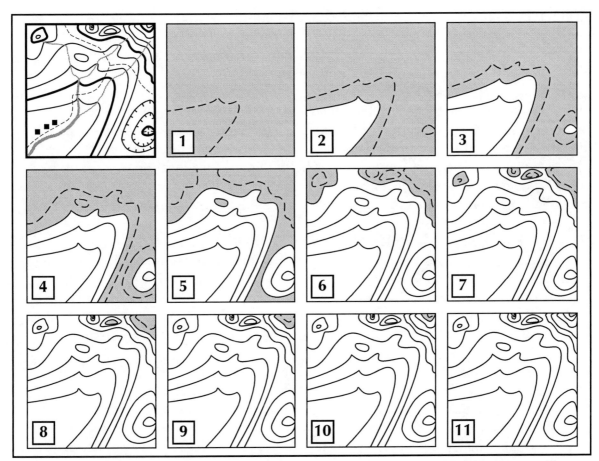

Here is an example of how a three-dimensional model goes together. The enlarged topographic map is at the top left. The sequence of steps 1–11 shows how each layer (light grey), beginning with the plywood base, is added and how it is marked with the outline of the next layer (dashed line). Notice how the shapes and placement of the cardboard pieces reflect the peaks toward the top of the map and the large depression (indicated by the ticks on the contour lines) in the lower right corner.

2. **Prepare the plywood on which you will build the model.** Use a sheet of carbon paper to trace the first contour line onto the board. Also draw the grid lines onto the board.

3. **Prepare and cut the cardboard.** Cut a sheet of carbon paper large enough to encompass the area within the first contour line. Place the carbon on a sheet of cardboard of a similar size. Put the topographic map on top and trace the first contour line onto the cardboard. Then trace the adjacent contour line onto the same piece of cardboard. Remove the topographic map. Cut the cardboard by following the first (outside) line. The second (inside) line will guide the placement of the next sheet of cardboard.

Now place carbon paper and the topographic map on a second sheet of cardboard and this time trace onto it the second contour line again and the third line. Cut the cardboard on the second line and use the third line as a guide to place the next sheet of cardboard.

Continue tracing onto and cutting the cardboard in the same fashion with lines three and four, four and five, five and six, etc. *Note that, as mentioned for the first contour line in step 1, the contour lines of a landform may split into several enclosed areas above a particular elevation. Just cut two (or more, as necessary) pieces of cardboard instead of one for each layer as you get to it. If there are depressions in the land, you will need to trace and cut not just the outline of the affected layers, but holes as well. For complex terrain, be sure to keep track of where each piece belongs after you cut it.*

4. **Place and glue the cardboard onto the plywood.** Place the first cardboard piece (with contour lines one and two) according to the first contour line that you traced onto the plywood. Place the second cardboard layer (with contour lines two and three) by fitting it within the number two contour line. Continue layering the cardboard in this way. Be sure to check the fit without glue each time before you apply glue and press the cardboard into place. You will soon see the shape of the hills or mountains begin to form.

A three-dimensional model of a village watershed in Thailand.

5. **Spread plaster over the cardboard model.** Use the plaster according to the instructions for the brand of plaster that you are using. Apply just enough to cover all the cardboard and to hide the steps between the layers. Allow it to dry overnight.

6. **Paint the model.** Consult the topographic map and paint rivers and other important features onto the 3-D map. Have local people indicate the different types of land use or vegetation on the model. Then paint the model using various colours to represent these areas.

7. **Make a legend.** Draw a legend somewhere along on the edge of the 3-D map to explain the colours that you used. Complete the legend by writing the title, a list of symbols, the date, and the number of the topographic map that you used. Also indicate the scale and draw a north arrow.

Store or display the model where it won't be damaged and where villagers have access to it. The model of the land is easier to understand than a topographic map and is therefore better for group discussions. Although a 3-D model is not useful for presenting to government officials, it is a good tool for interviewing groups of villagers for the purpose of drawing their knowledge onto flat topographical base maps. It is also useful for ongoing discussions about land issues amongst villagers.

A 3-D model can be used to

❖ *Discuss boundary conflicts between villages or between families*
❖ *Clarify information that villagers have given to be drawn on maps*
❖ *Discuss and develop a local land-use plan*
❖ *Discuss both internal and external threats to the village environment*

7.4 TABLE-TOP MAPPING: SKETCHING THEMATIC INFORMATION ONTO A TOPOGRAPHIC BASE MAP

Topographic maps make good base maps because they are full of reference information, such as the locations of mountains, ridges, rivers, and roads. Using this reference information and local knowledge of the land, you can draw thematic information straight onto the topographic map with no need for any tools or measurements.

For example, many communities use natural features such as the **height of land** (mountains and ridges separating one river system from another) to define their boundaries. It is easy to draw such a boundary on a topographic map by drawing a line that connects the summits and saddles and ridges. In many countries around the world in which traditional territories are large, communities draw their maps this way—they never actually survey the boundary on the ground.

Although it is better to do the drawing while walking on the land, it can easily be done in the house or office, and thus the term **table-top mapping**. Table-top mapping can be done with any kind of base map or remote sensing image.

Sometimes community members may draw directly onto a copy of the topographic map. However, in community-based mapping, villagers often draw a sketch map before they have learned to read topographic maps. Topographic maps can be confusing to villagers at first because they already have a complete and orderly map in their own minds, which doesn't look like the mess of lines on the topographic map. A sketch map on blank paper is a better reflection of their perception of the land.

The quickest way to then make a scale map is to simply transfer the information from the sketch map onto a topographic map. A sketch map of customary lands usually includes natural features such as rivers and mountains and rapids and rocks. These features also appear on a topographic map. By using these natural features as reference points, we can draw the information from the sketch map onto the topographic map. This approach is easiest in mountainous areas, where there are many well-defined reference features—such as mountains and a distinctive river pattern. It is virtually impossible in flat lowland areas.

Transferring information from sketch maps to a topographic map should always be done in consultation with the people who drew the sketch maps and know the land.

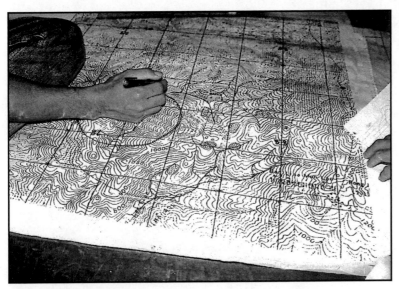

For each feature on a sketch map that we match to the topographic map, we can ask questions to verify the match and add information around it. With a little guidance, most villagers become excellent at reading maps. They have no problem understanding topographic maps, even at different scales, so long as specific features or landmarks are pointed out to help orient them.

Sketching local knowledge directly onto a topographic map.

WAYS TO HELP VILLAGERS READ A TOPOGRAPHIC MAP

☞ *Make a three-dimensional model with the villagers.*

☞ *Colour the lines on the topographic map (if you are using a black-and-white photo-copy) to highlight the different kinds—blue rivers, red roads, green vegetation lines, brown contour lines.*

☞ *Take the map outside and orient the map to the land to help them recognize how the map relates to the land. Have the villagers point out the peaks, rivers, or other major features that they know well.*

☞ *Naming the rivers is a good starting point. The river system of most places forms an orderly framework. Local people often use the river system for navigating over the land, so they are very familiar with the sequence, angles, and relative sizes of the rivers. Start at a place that you know, such as at the confluence of two major rivers. Follow the big river on the map, naming in order all the rivers that enter it from each side.*

☞ *Ask questions such as:* What is the first small river to enter the major one above the junction? What is the next river upstream? What do you call the peak at its headwaters? *Ask about the relative sizes of the rivers and where the headwaters are. See if it makes sense on the map. If the names on the base map are wrong, change them. Refer to the sketch map that the villagers made to help you think of questions.*

☞ *Take the map out to a viewpoint in the company of community elders. Orient the map and point out the major landmarks.*

Once people are accustomed to reading the topographic map, then all kinds of thematic information can be drawn onto it; locations can be determined just by reading the shape of the land. Use coloured pencils and symbols as you draw to help you keep different kinds of information from becoming confused with each other when you later draw the final map.

A thematic map drawn this way we could call a *scale sketch map*. It is a sketch given scale by fitting it to a topographic map. This kind of map is more credible than a simple sketch map. Indeed, government agencies often use this technique to draw the boundaries of their land-use designations. Thus, for initial dialogue or negotiations about land-use conflicts between local people and outsiders, this kind of scale sketch map is sufficiently accurate.

8 USING A TOPOGRAPHIC MAP WITH A COMPASS

What you will learn in this section:

➢ *What a compass is and how it works*
➢ *When and how to adjust for magnetic declination*
➢ *How metal objects can affect your use of the compass*
➢ *How to hold a compass*
➢ *How to orient a map using a compass*
➢ *How to measure the direction (bearing) to a feature on the land.*
➢ *How to plot a bearing onto a map*
➢ *How to take a bearing from a map*
➢ *How to use a compass to locate your position on a map*

8.1   ABOUT COMPASSES

A compass is a tool used in the field to measure *azimuth*—the amount of angle that a given direction is from magnetic north. A typical compass contains a magnetized 'needle' (a small, thin strip of metal) mounted so that it can respond freely to the Earth's magnetism. Using a compass is a quick and accurate means of establishing direction, because the needle aligns itself with the magnetic field of the Earth so that the coloured (usually red or black) or pointed end of the needle points toward magnetic north.

8.1.1   TYPES OF COMPASSES

Compasses are made in many forms. The simplest kind of compass consists simply of a magnetic needle that is free to rotate over a card that is labelled clockwise with the directions—N, NE, E, SE, S, SW, W, NW—all enclosed in a container shaped like a wristwatch, but without a wrist strap. This type of compass is not very useful for making maps.

Another common type of compass consists of a magnetic needle that is free to rotate within an oil-filled *capsule*, which in turn is mounted on a *base plate*, where the capsule can be rotated by the user. The capsule has a north arrow and parallel lines marked on its bottom, below the nee-

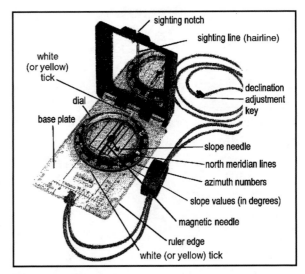
The parts of a compass.

dle. The circumference of the capsule is usually marked from 0° (north) clockwise past 90° (east), 180° (south), and 270° (west), to 0° again (which is also 360°), with small ticks to indicate every 2° or 5°. These compasses are held in the hand and the sights are very simple, so there can easily be small errors. The transparent base plate usually found on this style of compass makes it very easy to transfer bearings directly to the map.

A more sophisticated version has the same type of base plate plus a sighting mirror to make it easier and more accurate to sight on a target (Silva—whose compasses are now labelled 'Nexus' in North America—and Suunto are common brands). Field compasses like these ones are used by foresters, parks managers, researchers, land-use planners and recreationists all over the world because they are relatively inexpensive, reasonably accurate for their purposes, and fast to use. These compasses are the best ones to use for community-based mapping.

Land surveyors and road engineers use a more complex tool called a **theodolite** that consists of a compass and a level mounted on a tripod. Although more accurate, a theodolite is slower to use and considerably more expensive.

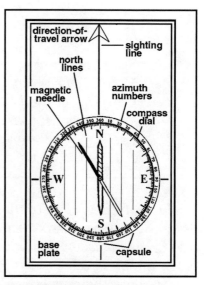

Each of the small marks on this compass dial is equivalent to 2°.

Note that it is best to use a compass made for use in your area or one designed for universal use. Compasses made for northern regions have their needles weighted differently than the ones intended for southern latitudes. If you use a compass meant for the north near the equator or in the south, the needle will tend to dip into the card and get stuck instead of rotating freely on the horizontal.

8.1.2 DECLINATION

Remember that the compass shows **magnetic north**, which is not the same as either the grid north or the true north shown on maps (review subsection 2.2.3 if necessary). As you probably recall, **grid north** is the north used by the particular coordinate system that your map is made on, whereas **true north** corresponds to the Earth's axis. Confusing? Don't worry. A topographic map tells you how much the declination is, and a good compass should have a way to adjust for magnetic declination. Once you adjust the compass for your area, you don't have to think about declination again (unless your area is relatively close to either magnetic pole).

Use the small key that comes with your compass to adjust it to the declination for your area.

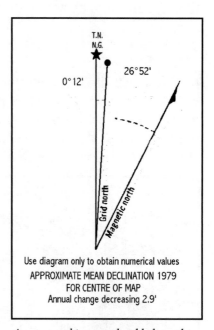

A topographic map should show the declination for the area at a particular date, including whether it's east or west, and the rate of change and whether it's eastward or westward. Calculate the declination for the present year and adjust your compass accordingly.

HOW TO CALCULATE DECLINATION FOR THE PRESENT YEAR

1. *Subtract the year the map was made from the present year.*
2. *Multiply this number of years by the amount of change each year.*
3. *Depending on the direction of the change, either add this number to the value given for magnetic declination on the map or subtract it. For example, if the declination when the map was made was to the east and is changing to the east each year, then add. If the declination was east and is changing to the west, then subtract.*

Look in the margin of the topographic map to find an arrow that tells you the difference in angle between magnetic north, grid north, and true north for the year in which the map was made. Magnetic north continues to change in relation to true north by a few minutes every year. We have to calculate the declination for the present year.

To adjust the magnetic declination setting of the compass (the Silva or Suunto types), there is a small (special non-magnetic) screwdriver on the string, which you use to turn a tiny screw in the rim of the compass dial. Be sure to take into account that declination can be either east or west (see the compass diagram).

If you have a simpler compass—and the declination is over about 5° or so—you need to take declination into account by adding (or subtracting) on every measurement. A simpler solution is to carefully draw magnetic-north lines across the map at intervals of 4 or 5 cm. Because there is a difference between magnetic north and true north, and because magnetic north changes slightly over time, it is important that true north is also shown on the map.

8.1.3   LOCAL ATTRACTION

An important point to remember when you are using a compass is that it is sensitive to many metallic or magnetic items. The compass needle can be deflected by electric cables, cars, guns, flashlights, screwdrivers, knives, some coins—even staples in the map (and on rare occasions, a buried object, or a metallic mineral deposit). And really strong magnets (for instance, in radio speakers) can actually damage the compass (by demagnetizing the needle) if you bring it too close.

Don't use a compass within 3 m of a wire fence or a vehicle, or it may affect the accuracy. If you are carrying a gun or machete, it may be sufficient to put it on the ground behind you, but if you notice that it is still affecting the compass (if your bearings seem to be 'off'), then move it 3 m away.

Don't use a compass near large metal objects.

8.2 HOW TO USE A COMPASS

8.2.1 HOLDING THE COMPASS

In the following instructions, we will consider the hand-held field compass with a sighting mirror (Silva or Suunto types) because that is the type that you will probably be using. Whenever you sight with a compass, hold it out at roughly arm's length, at about eye level, with the mirror adjusted so that you can clearly see the compass capsule in the centre. (The mirror should be not quite vertical, but leaning toward you a little). Look at the feature through the slot at the top, and line up the reflections of the white or yellow ticks on the base plate with the hairline on the mirror. Make sure that the compass is level and the needle can move. (Near the equator the needle tends to pull down into the compass plate if you use a model designed for the north.) Check to see it is free to move by lining it up, with your eye to the feature, and then rotating the compass a little in each direction to see if the needle stays in the same place as the compass moves around it (as it should).

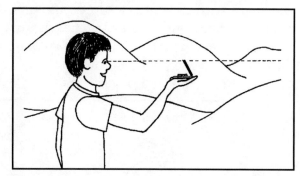

Hold the compass just below eye level. Tilt the mirror to see the face of the compass. Then sight the object while checking that the alignment marks line up.

Especially when surveying, always hold the compass with the same hand and in as close to the same position as practical each time, and look through the sight with the same eye. Hold it level, and be especially aware to look straight at the compass instead of at an angle, so that you see the needle in its true relationship to the markings on the capsule base.

If the compass does not have a mirror, you will need to hold it a little lower and closer to you in order to see where the needle points. Also be sure that the **direction-of-travel arrow** (through the centre of the base plate and aligned with the sights) points directly toward the feature, and is not skewed to one side or the other.

8.2.2 READING THE BEARING

A **bearing** is the angle of a line in relation to the direction of magnetic north. To use the compass to find the bearing to some feature on the land, such as a mountain peak, point the compass sight in the direction of the peak. Turn the compass dial or capsule until the north arrow on its base aligns with the magnetic needle. (Make sure that it's the right way around and not reversed!) Bring the compass close and read the bearing at the indicator mark.

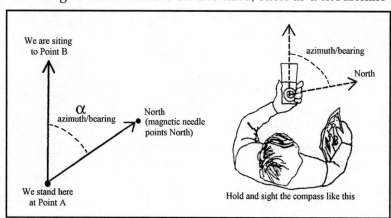

Reading a compass.

8.2.3 FINDING THE DIRECTION TO A FEATURE USING A BEARING

Conversely, if you know the bearing (say, you got it from your notes or someone told you) and you want to find out where it is on the land, you want to do the reverse of what it says in subsection 8.2.2: Turn the compass dial until the bearing number is at the base-plate tick. Then turn the whole compass, and your body with it, until the needle aligns with the north arrow on the base of the capsule. For example, if you want to find out where east is, turn the compass dial until 90° (east) is at the tick on the compass base plate. Then turn the whole compass until the north end of the needle lines up with the north end of the arrow on the compass base plate. (Check that the compass is at right angles to your body and that it is level.) You and the compass now point east.

Use the compass to sight precisely to a feature, such as a mountain. Turn the compass dial until the capsule arrow aligns with the needle. Read the bearing off the dial.

8.3 HOW TO USE A COMPASS WITH A TOPOGRAPHIC MAP

8.3.1 ORIENTING THE MAP USING A COMPASS

Turn the compass dial to 0°/360° north. Find the north arrow on the map. If you've adjusted the compass for declination, then use the map's true north arrow; if not, use the magnetic north arrow (check the declination diagram). If the shaft of the arrow is short, draw it longer with a ruler or the edge of the compass. Place the compass base plate so that the long edge is lined up neatly beside the map's north arrow's shaft or a north line. Then turn the map and compass together until the magnetic needle in the compass aligns with the north arrow on the bottom of the compass capsule. (The needle should point in the same direction as is shown for north on the map.)

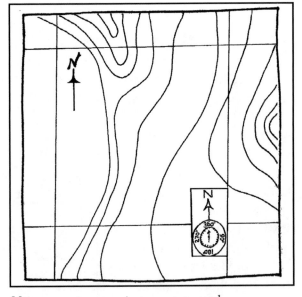

Using a compass to orient a map to north.

8.3.2 PLOTTING A BEARING ONTO THE MAP

If you've taken a bearing to a feature on the ground in front of you, and you want to find in what direction that feature is on the map, you have to *plot* the bearing on the map. To do so, keep the bearing number at the tick on the compass base plate. Because you will be using the compass as a protractor only, ignore the magnetic needle in this procedure. Place the compass on the map and turn the whole compass until the parallel lines on the base of the compass capsule align with the grid north lines on the map (or, if you are using a simple compass without declination adjustment and you have drawn magnetic north lines on the map, align with them). Move the compass along the grid line until either long edge of the compass base plate touches the point that you know on the map (that is, the point on the map that represents the place where you stood as you took the bearing). The bearing follows the edge of the compass base plate.

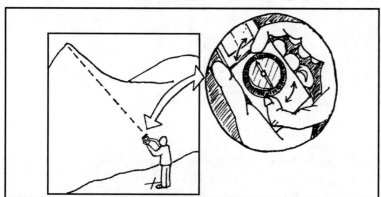 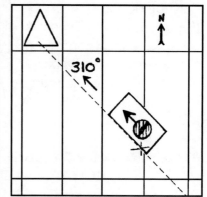

To plot a bearing on the map, first set the bearing on the compass (by sighting in the field, or from notes). Then put a long edge of the compass on a known point on the map (for example, your location or a sighted feature). Finally, pivot the whole compass as a unit around this point until the black lines on the capsule base align with a north line on the map. You can now draw in the bearing (or mark a second point at some location along it).

If you need to, use a ruler to extend the line to connect the feature and your own location. Note that it is usually best to draw bearings lightly on the map, to make erasing easy, unless you need to record them graphically in addition to the locations of the features themselves.

As is explained in detail in the following sections, a single bearing can give you the direction only—to know the location of the feature requires either another bearing from a different location, a measurement (or estimate) of distance along the bearing, or a careful comparison of the shape of the feature with the markings on the map so that you can identify it.

These community mappers took a bearing to a nearby island and are now locating it on their map.

8.3.3 TAKING A BEARING FROM THE MAP

If you want to take a bearing from the map, place the compass on the map, with either long edge of the base-plate lined up with the two points that you want to find the bearing between. The compass should point in the direction that you are looking or are planning to go (and not the opposite direction). Ignore the magnetic needle.

Turn the compass dial until the parallel lines on the capsule base are parallel to a north grid line (or with magnetic north lines, as described in sub-section 2.2.3). Make sure that the north arrow on the compass capsule points toward the top (north end) of the map. Read the bearing at the tick at the top of the compass plate and record it.

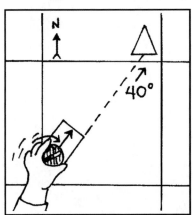

Taking a bearing from the map to use in the field.

8.3.4 USING BEARINGS TO FIND POSITIONS ON A MAP—RESECTION, INTERSECTION, AND TRIANGULATION

When you have oriented the map and identified some ground features, you may still not be sure exactly where you are on the map. Perhaps all the ground features are too far away to give you a good indication. As was mentioned in subsection 8.3.2, one problem in using a compass to locate things on the map is that you only know the direction, and not the distance. You could be located anywhere along the bearing line to the landmark that you are sighting on. But, with the three techniques explained below, you don't need to measure the distance. You can determine a location by taking bearings to (or from) two known points.

Note: Be sure to check 'Some Points Common to Resection, Intersection, and Triangulation' at the end of this subsection.

Resection

Stand where you can see two landmarks that you have identified on the map. Take bearings to both points and plot the bearings on the map. Draw a line in pencil at the measured angle (bearing) from each of the landmarks. Where the lines cross is your position on the map. This procedure is called **resection**.

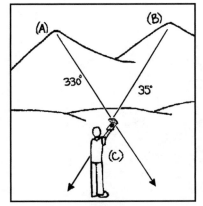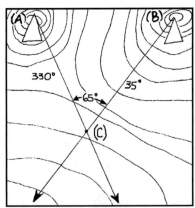

Resection: Find bearings to two features that you can identify on the map (A and B), and then plot the bearings. Your location is where the lines cross, at C.

It the two features were not very close to 90° apart, or if you have other reasons to check the precision of your calculated position, you can take a third bearing to another feature (see the description of triangulation in this subsection).

However, if you are standing on a linear feature—such as a river or a road—that you have already identified on the

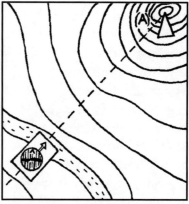

If you are standing at a river that you have identified on the map, then take a bearing to a mountain that you have identified on the map. Where the line crosses the river is your location.

map, then you can do resection with only one bearing. Take one bearing to, for example, a mountain whose location you also know on the map. Where the bearing line crosses the river or road is your exact location. (It is best if the bearing is at approximately 90° to the linear feature.) If the river or road curves, you may find that the bearing intersects twice or more and you will need to decide which position is the correct one.

Intersection

Intersection is the opposite of resection, in that you have to walk to different known locations and take compass bearings from each of these locations to the same feature, for example a sacred tree, that you don't know the location of. This technique is commonly used while doing a compass survey (see chapter 9), because the exact locations

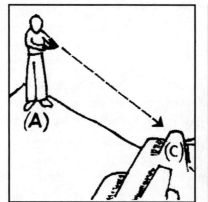

Take bearings from two separate known locations, A and B, in order to find the location of an unknown location, C.

of the stations are known and you have to walk to each station anyway.

Triangulation

You may have taken two bearings from one unknown location to two known locations (resection) or from two known locations to an unknown location (intersection). If you know a third location, then you can take a third bearing to verify the first two. (Where possible, choose a third feature midway between them, but in the opposite direction. If the direction of the third bearing is close to that of the first or second, it will confirm—or contradict—only that bearing and not the other one.) Usually there is a small error and the three bearing lines won't cross at one point. Rather, they'll form a small triangle. You can then consider your position to be in the centre of the triangle. When is the triangle so big that you know that there is a mistake? It depends on the scale of the map, but a good guideline is that each of the triangle's sides should be shorter than 1 cm.

 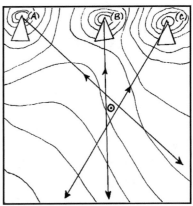

Small triangle or large, it's always good to check with the contour lines, and the river pattern, to see if your position makes sense. If you are on a ridge, and

Triangulation. Take bearings to three or more known points from an unknown location. Plot the bearings on a map and the lines will cross to form a small triangle. The centre of the triangle is your position.

you find that the bearing lines cross at a river, something is wrong. Maybe the mountains that you are sighting on are not the ones that you think they are on map. Maybe you held the compass wrong and made a mistake on the bearing. Maybe the declination is set wrong on the compass. Maybe the map is not precise enough, for example, when an entire mountain peak is shown as a circle 2 cm in diameter. Still puzzled? Consider taking a fourth bearing, if possible.

Some Points Common to Resection, Intersection, and Triangulation

The techniques of resection and intersection depend on knowing a minimum of two locations on the map and on the land (triangulation requires three or more). Thus, when used with a topographic map, these techniques are the most useful in a place with an adequate view of hills, peaks, or other features that are easily identified.

The optimum accuracy is when the two known points are 90° apart. Closer or farther than that adds error. In practice, aim for a figure between 60° and 120°. (If the bearings are too close—or they are about 180° apart—you will confirm the direction, but not the location along it.)

The more precise the sighting point on each feature, the more accurate your bearings will be. If you are sighting on a rounded mountaintop, you should choose the very highest point, or the point in the centre. Depending on the contour interval, the exact point that you are sighting on may be difficult to pinpoint on the map. There is nothing you can do about that (apart from choosing another feature). Just keep it in mind that it can be a source of error.

As you might guess, if you are locating many points using intersection (and thus have to find two places to take bearings from for every point), you could be doing a lot of walking! If you have a lot of features to map this way, sight on as many as practical from each viewpoint, carefully recording on a sketch which feature each bearing is for. Then complete the intersection for each feature at the second viewpoint from which you take a bearing on it. This technique can work well when you are mapping a small valley from the surrounding hills.

9 HOW TO DO A COMPASS SURVEY

What you will learn in this section:
- ➤ *What a traverse is*
- ➤ *How to survey linear and closed traverses*
- ➤ *How to measure distances while surveying*
- ➤ *How to compensate for slope*
- ➤ *Four techniques for filling in details along a traverse*
- ➤ *How to record a traverse in a notebook*
- ➤ *How to plot a map of a traverse from field notes*
- ➤ *How to check the accuracy of a closed traverse*

9.1 WHAT IS A COMPASS SURVEY (TRAVERSE)?

A *compass survey* (also known as a *traverse*) involves walking along a chosen route and measuring the distance that you walk with a metre tape and the directions that you walk in with a compass. The bearings and distances are recorded in a notebook and then used to *plot* (draw) a map on graph paper.

A traverse always implies surveying along a line. It could be a straight line or it could be a line that follows a winding trail. Record what you see on either side of the line and you can draw a reasonably accurate map, but only of that line and what is adjacent to it. To make a map of an area, you need to survey several lines, or else fill in details by sketching.

A compass survey (traverse) is the simplest way to make a scale map without using a topographic map. The tools are simple, but a compass survey is, in fact, more accurate than a survey done with a non-differential GPS receiver (read about GPS surveys in chapter 10). Because compass surveys take time to do, they are not used for mapping a whole territory while doing community-based mapping. They are used selectively, where we want to accurately record a lot of information at a large scale. For example, a compass survey is useful for mapping houses along a road, or the perimeters of farm fields in order to measure their areas.

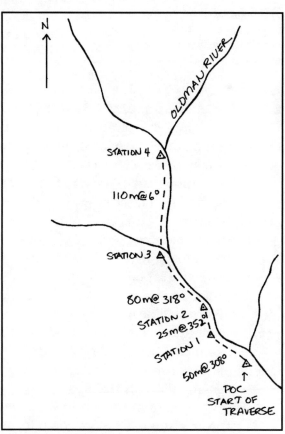

A compass survey, or traverse, entails walking a route and measuring bearings and distances along the way.

9.2 STEPS TO PLAN A TRAVERSE

9.2.1 ORGANIZE THE SURVEY TEAM

The minimum team to do a compass survey is two people: one to hold each end of the metre tape. One of the two also takes compass bearings. Although not essential,

it is easier with a third person taking notes. Also, it is critical that at least one villager on the team knows the land well. This resource person is likely an elder. If he or she is not interested in doing technical surveying or note-taking, then the resource person becomes a fourth person on the team.

A priority in community mapping is maximum participation. Although this goal may not mean that every villager needs to participate in the compass survey, the more resource persons the better. If there are many people, they can divide into two, three, or four groups, each surveying a different route.

A person (or several people) in the community should be selected to coordinate the survey and take responsibility for the data. That person (or group) must have the interest and capacity to understand the technical principles of surveying in order to coordinate the efforts of all the villagers who would like to participate. If the community already selected a mapping team (see section 2.7), then the coordinator might be a person from the team (or several people could share the job).

Traversing can be done with two people, but in participatory mapping there are usually many more villagers who want to help.

9.2.2 CHOOSE FROM FOUR BASIC TRAVERSE PATTERNS

There are four basic traverse patterns. Depending on your objectives, you may choose to use one or more of them for your mapping project.

❖ *Linear traverse:* A survey that simply follows any desired line; for instance, a winding road or a river. The line could also be a straight line through the forest. This pattern is commonly used in community-based mapping, because a lot of important features are located near roads and rivers.

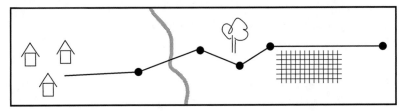

A linear traverse goes in essentially one direction, though it can be straight or it might zig-zag.

❖ *Boundary traverse* or *closed traverse:* A survey around the perimeter of an area—for instance, following a land boundary all the way around, back to the starting point. If the area that you are surveying is rectangular, then you will only need to measure four sides and take four bearings. However, the shape of the area is usually more complicated. You will know if you have made an accurate survey when you draw the map: the lines should meet neatly at the same place, without leaving a space and without crossing. (Because you go all the way

around the area, coming back
to where you started and resulting
in a closed figure on the map,
a boundary traverse is often called
a *closed traverse*.)

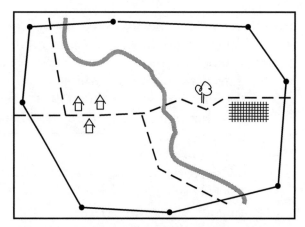

A boundary traverse is used
to make a simple, accurate map
of an area that has some kind of a
boundary, not necessarily a marked
one. The area could be a whole
traditional territory, a farm, a field,
a land parcel, an area of logging,
an area to be flooded by a dam,
an area to be made into a plantation, and so on.

When you do a closed or boundary traverse, you should end up back at your starting point.

A boundary traverse is the only pattern from which you can calculate the number of hectares within the area traversed.

❖ *Grid traverse:* A survey that consists of several straight linear traverses parallel to each other. A grid traverse is used when you want to map a lot of detailed information over an area in a systematic manner. Begin this technique by starting in one corner of the area and following the same compass bearing (roughly parallel to one edge of the area) as you travel across the area. Then change the bearing 90° to go across to the next line, which is chosen to be offset at a convenient distance to the first. How far depends on visibility—you need to be able to see the level of detail that you want without overlooking anything. Then follow a bearing 180° from the first to make the second line parallel to the first.

If you make the lines 50 m apart and draw a map of what you see 25 m on each side of your traverse line, then your survey will cover the whole area. Any point that you pass on the traverse line can be located precisely on the map. For instance, if a road crosses the land, if you plot the location of every point at which the traverse line crosses the road, you will be able to accurately draw the road onto the map. Foresters often use a grid pattern when they want to make an inventory (count the trees of each type of interest in the forest).

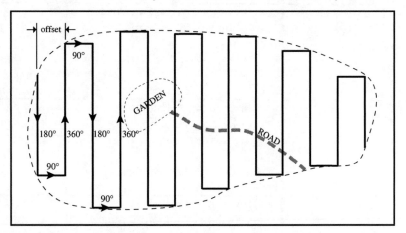

A grid traverse usually records the most detail, and in a systematic way, but it takes the most time. Because it involves walking

A grid traverse is the most systematic way of filling in the detail of an area. You can see that you can easily map the road or river because you know exactly where the traverse line crosses it at several points. With an offset (spacing) of 50 m between successive 180° and 360° sections, you will usually (depending on visibility) not miss much or do too much surveying.

in straight lines across the land, this kind of survey may not be practical, depending on the topography and the ground cover.

❖ *Radial traverse:* Made by doing linear traverses starting at one central place. This pattern, like the grid traverse, is also good for mapping details of what is inside a given area. You might want to do a radial traverse if the land is like that of many communities, where the village is in the centre and there

are many trails going out to the gardens and the forest. You could traverse along the trails, measuring distances and bearings, and recording both natural landmarks and community landmarks along the way.

This method is especially good if most of the things that you want to

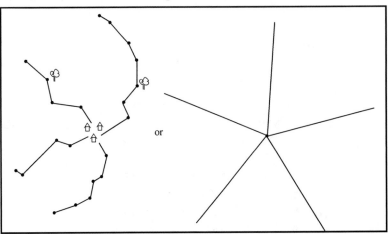

A radial traverse is a good pattern to use if most of the details you want to map are near roads or trails or rivers, or if the land is easy to walk.

map are near the trails or roads and the terrain or vegetation makes off-trail walking difficult. But remember that between the trails there may be significant

CHOOSING A TRAVERSE

TYPE OF TRAVERSE	CHARACTERISTICS	WHAT TO USE FOR
LINEAR	☞ The simplest, most basic traverse ☞ Gathers information along a single path only	☞ A cross section of an area, such as mapping how land use or vegetation varies with elevation ☞ A single road, trail, or river
BOUNDARY (CLOSED)	☞ A linear traverse that returns to its starting point to form a closed loop ☞ Goes all the way around an area	☞ To survey the boundary or perimeter of an area ☞ A good starting point if the area to be mapped is of manageable size ☞ Areas whose sizes you want to calculate
GRID	☞ A linear traverse that folds back on itself repeatedly ☞ The most thorough traverse ☞ Covers an area in detail ☞ The most time-consuming ☞ Requires walking in straight lines	☞ Mapping all the features in an area ☞ Especially useful if there is no topographic map available ☞ Land that is easy to walk across
RADIAL	☞ Consists of several linear traverses with a common starting point ☞ Can follow trails and roads ☞ Can cut across the land in straight lines (if practical)	☞ A good second choice for an area if a grid traverse is too much work and the land is walkable or there are trails ☞ Places where most or all of the features to be mapped are adjacent to roads, trails, or waterways that branch out from a common place

features that you will miss by doing this kind of traverse (though you can make a point of recording them using other techniques if you know where to find them).

With four kinds of traverses to choose from, how can you decide which kind to use? Well, it all depends on your mapping situation. Take into account these factors when deciding:

❖ *What kind of information you want to map*
❖ *How much detail you want to map*
❖ *How the important features are distributed*
❖ *The size and shape of the land*
❖ *How much time you have*
❖ *Are there a lot of roads or trails?*
❖ *Is it easy is to walk across the land in a straight line?*

In practice you will quite likely use not just one but two or more kinds of traverse for your mapping project. When combining traverse patterns, obviously it is important to join the surveys at a known station.

If you cannot get a topographic map for the area, then any of these traverses can be used to make a map by hand. Of course it won't be filled with as much detail as if it was based on a topographic map, which shows the shape of the land as well as the road and river networks.

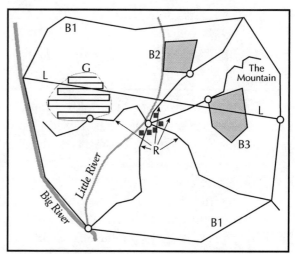

In this simplified example, a boundary traverse (B1) of the traditional lands was done first. Then the mappers did a radial traverse (R) along the trails that go through the centre of the village. The two garden areas (light grey) were selected for boundary traverses (B2, B3) so that they could be correctly drawn. An area on the left was known to contain a lot of sacred sites, so a grid traverse (G) was performed. Finally, so that they could draw a cross-section of land uses, the villagers did a linear traverse (L) across their land.

9.2.3   DECIDE WHAT INFORMATION TO RECORD

If the team is going to take the time to walk the whole survey, it should record as much information as possible. The time available is the only limit to how much information can be recorded while surveying. Before going out into the field, discuss among the survey team what kinds of information they expect to observe and record. Particularly if the survey team breaks into several groups, it is important that each group collects roughly the same kinds and amount of information. (Each should also have the same criteria for what distinguishes, for example, one forest type from another and how small a feature of each kind to record, and they should all use the same set of symbols— see section 4.2.5.) Some types of information that could be recorded are

❖ *Land ownership*
❖ *Land use*
❖ *Historical places*
❖ *Local names of places*
❖ *Forest products*
❖ *Gravesites*
❖ *Forest type*

9.2.4　PREPARE A NOTEBOOK OR FORMS TO RECORD THE DATA

Before going to the field, prepare either a notebook or sheets of paper for taking notes. Prepare a space to write the name of the village, the date, and who did the survey. Then draw seven columns like this:

Station #	Front Bearing	Back Bearing	Distance (metres)	Slope (degrees)	Horizontal Distance	Notes

The easiest kind of notebook to manage in the field has sheets that are 10 cm × 15 cm and can be opened flat. Divide the left-hand page into columns, and use the right-hand page to draw a sketch of the traverse. The advantage of a notebook is that it is small, but the disadvantage is that it takes time to prepare the columns.

You could also prepare your own form for recording data. Take a sheet of paper and use a ruler (or computer) to set up the columns on the left of the paper. Cut and glue a piece of graph paper to the right of the columns to use for sketching. The advantage of using forms is that you can photocopy as many as you need.

Clip the forms onto a board for use in the field. (With a little bit of extra work, you could even bind or staple sheets together inside a cover to make your own survey notebook.)

Read more about taking survey notes in subsection 9.4.3. For an example of a compass survey form, see appendix A.

9.2.5　SELECT A STARTING POINT AND PLAN THE SURVEY ROUTE

Carefully select the place to start the survey. It should be a place that can be marked and described precisely so that a person can find the exact place again. It should be marked permanently by some recognizable feature—if necessary, you can mark the location yourself with a post. Ideally, it can be found on a topographic map so that you can know where this survey is on the topographic map. A good starting place might be a large rock, or a post in the middle of the village, or a river junction or road junction. (Avoid choosing a feature that might shift with time, such as a field corner or the junction of two rivers that are quickly eroding their banks.) In surveyor's language, the starting point is sometimes called the *point of commencement* or *POC*.

Use a sketch map to plan the survey route. If there is more than one survey team, then it is critical that each one understands where to start and finish its part of the survey.

Select a natural landmark or make an immovable marker as the staring point of your survey.

9.3 MEASURING DISTANCES AND DIRECTION

9.3.1 MEASURING DIRECTION

To measure direction on the land, use a compass to take bearings (covered in subsection 8.2.3), but in surveying you will be taking bearings to a partner, instead of to distant mountain peaks.

9.3.2 MEASURING DISTANCES

There are a variety of tools for measuring distances. The least expensive and most readily available is a regular rolled nylon or fibreglass metre tape that extends to 30 m or 50 m. Sometimes the tape has metres on one side and feet on the other. Make sure to read the correct side. In the west it was common to use a *survey 'chain'* made of nylon rope and marked by a metal clip at every one tenth of a metre. If you need to, you can make a metre tape out of a piece of rope, but it can be difficult to mark the

An effective way to measure distances is with a 50 m tape.

ESTIMATING DISTANCES

A quick method for estimating distances on the ground is pacing. Figure out your natural pace by measuring 100 m along a road. Count the number of steps it takes you to walk that distance. Count every left foot (or every right one), rather than every foot—it makes counting simpler. Do it three times and calculate the average number of paces (50 would be nice!). Then measure off 100 m on a medium slope and calculate an average number of paces for both uphill and downhill directions. Now try a steep slope. See how the number of steps changes. What happens if you travel along a trail instead? Through the forest? With a heavy pack on your back? To maintain accuracy, you must compensate for shortened or lengthened steps when climbing a hill, descending

a hill, or avoiding obstacles. On a very steep hill, every 100 m you should estimate the average angle of the slope in degrees and calculate the horizontal distance (described in subsection 9.3.3).

It is also possible to buy inexpensive pedometers (they attach to a leg or pants pocket) that measure how far you walk.

There are other ways to estimate long distances. For instance, you could use the odometer of a truck. Or you could time how long it takes to travel a kilometre by riverboat, or how long it takes you to walk a kilometre, and then measure distances by how long it takes you to travel them. Of course, whenever you measure distance by how long it takes to get there, you need to make sure that your average speed matches that over the known kilometre—and remember to subtract all your breaks.

metres on it. For instance, knots get caught on things and make the rope difficult to pull, whereas pen or paint markings often become erased. Make sure it is a kind of rope that doesn't stretch and is convenient to work with.

When surveying, the person in front carries the '0' end of the tape, chain, or rope. The number of metres is read from the tape by the person following behind.

Another tool for measuring distance is called a **hip chain**. It is a small plastic box, worn on a belt, that houses a roll of thread. The thread is pulled through a small counter wheel, calibrated so that a certain number of rotations is equivalent to a certain distance. The distance in metres and tenths of metres can be read directly from the counter wheel display. The advantages of a hip chain are that your hands are free, and it is easy to read and fast to use. One disadvantage is that it is less precise than a tape or regular chain, because the thread is fine and easily broken and so cannot be pulled straight and tight. Another disadvantage is that it requires thread refills.

9.3.3 WHAT TO DO WHEN MEASURING DISTANCE ON A SLOPE

The distance that we measure when two people hold the metre tape parallel with the ground is called the *slope distance*. Maps are drawn using *horizontal distance*, or the flat distance. The measured slope distance is always longer than the actual horizontal distance. The steeper the slope, the greater the difference between the two. You can see that if you draw a map using the slope distances, exactly as they are measured in the field, then all the distances that you measure on the traverse will be somewhat too long, and the map will be distorted. If you encounter slope on the traverse, you can either measure or calculate the corresponding horizontal distance.

Measure the Horizontal Distance While Surveying

Measure the horizontal distance by always holding the metre tape horizontal, rather than following the slope. To do so, the person at the top of the slope might need to hold the metre tape at their feet, while the person at the bottom of the slope holds the tape above his or her head. If the slope is very steep and or long, then it is not possible to hold the tape horizontal for the full distance to be measured. In this case you need to measure longer distances as shorter lengths or 'steps.'

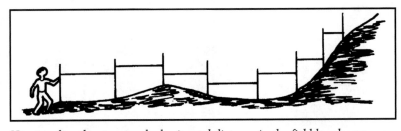

You can directly measure the horizontal distance in the field by always holding the measuring tape horizontal, regardless of the slope.

Calculate the Horizontal Distance After Measuring the Angle of the Slope

Another way to get the horizontal distance is, for each leg of your survey, to measure the average angle of slope at the same time as you measure the slope distance. Later, when processing the data, you

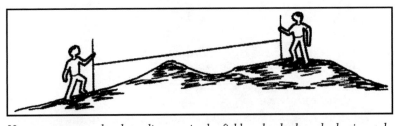

You can measure the slope distance in the field and calculate the horizontal distance later.

will calculate the horizontal distance from the slope in degrees (or percent) and the slope distance. There are a variety of tools for measuring the slope angle. Here are three of them:

❖ *Clinometer*: The best instrument to use to measure slope is a clinometer. Sight at a target through the clinometer to directly read the numbers for either the percent or degree of slope. Clinometers are relatively expensive. Depending on the level of accuracy you need, which depends on the scale and purpose of the map, you could instead use the tools below for estimating the slope.

❖ *Slope needle on a compass*: The type of compass with a sighting mirror often has a slope needle. Turn the compass dial so that 'west' is at the top, near the mirror. Then turn the whole compass so that it stands on edge and look for the small numbers (usually red) on the inside of the dial, with 0° at the very bottom. When you hold the compass level while it is on edge, the slope needle (usually black) hangs at 0°. The more you slope the compass in either direction, the higher the number that the needle will point at.

❖ *Hand-made slope guide*: You can also use a protractor or make yourself a chart from 0 to 90°, which you can hold up at eye level to estimate the slope. Use a level or a plumb bob—or even an erect tree—to align the chart to the vertical.

Estimating the slope degrees using the slope needle on a field compass.

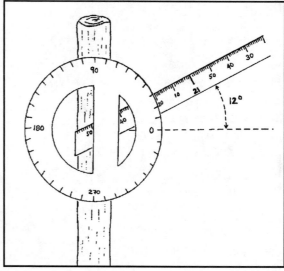

Estimating the slope with a protractor.

Calculate Horizontal Distance from Slope Distance

The steeper the slope, the greater the difference between the slope distance and the horizontal distance (which is shorter). If the slope is under about 5° or 10%, the slope distance almost equals the horizontal distance, so you don't need to do any calculations. If the slope is greater, then use this equation to calculate the horizontal distance:

horizontal distance = slope distance × cosine of the slope angle

Regardless whether you measured the slope angle in degrees or percent, you can find the cosine with a calculator or look it up in a table (see appendix B). There are also detailed commercial surveying slope tables available that don't require any calculations at all to get horizontal distance from slope distance.

9.4 DOING A TRAVERSE

9.4.1 PUTTING THE STEPS TOGETHER

Start the traverse at a visible landmark that is easy to find again. Each time you change directions, you need to take another compass bearing. A point on the traverse from which you take a bearing is called a *station*. Measure the distance between each pair of stations and record it in your notebook. Each station needs to be marked so that the compass person can find it. A suitable mark might be a broken branch, a stick in the ground, a small blaze on a tree, a small pile of stones, or paint on a tree. As much as possible, make stations at permanent features (river junction, house, big rock, etc.) so that it will be easier to find the station again during subsequent surveying.

You need a minimum of two people to do an accurate traverse. At the start of the traverse, the first person (leader) takes the '0' end of the metre tape and walks ahead to station 1. The second person is the compass person. He or she stands at the starting point (or *point of commencement*, POC) and holds the roll end of the metre tape as the leader pulls the tape out. When the leader arrives at station 1, he or she makes a mark, stands still and holds the tape firmly with the '0' exactly at the mark. The compass person pulls the tape tight, reads the distance, and records it in a notebook. Then he or she takes a compass bearing to the leader (who is at station 1) and records this measurement as the *front bearing*.

Then the leader walks to station 2, makes a new mark, stands still, and holds the tape tight. Meanwhile, as the leader

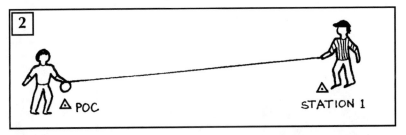

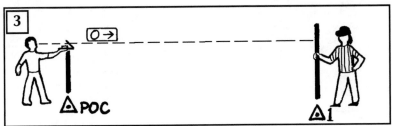

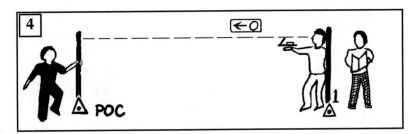

A traverse usually involves at least two people working together. The leader walks ahead to the first (or next) station with the free end of the metre tape (1). The compass person then takes and records both a distance measurement (2) and a bearing (3). (A pole can be held vertically right at the station as a sighting point for more-accurate bearings.) The compass person then goes to join the leader. If there is a third person on the team, he or she can remain at the POC (or previous station) so that the compass person can take a back bearing (4). The process is then repeated for each successive station. If there is a fourth person, he or she can accompany the compass person and take additional notes and make sketches (4).

moves toward station 2, the compass person walks to station 1 and waits until the leader arrives at station 2, allowing enough slack in the tape for the leader to progress. Then the compass person pulls the tape tight, reads the distance, takes a bearing, and records the distance and bearing in the notebook.

If there is a third person, then that person waits temporarily at the POC (or previous station) so that the compass person can take a compass bearing from station 1 (or current station) back to the previous one. This **back bearing** is used as a check for the accuracy of the original bearing (from which it should differ by 180°).

If the leader arrives at a river, road, or other important feature before reaching the next planned station, then he or she stops, marks the point as a new station, and signals for the compass person to measure the distance from the last station.

An additional person could walk with the compass person and take notes and sketch everything of interest along the survey. It is helpful to to take notes about natural features at the stations, as well as other information that is not visible but is local knowledge available to the survey team—land ownership, for instance.

9.4.2   MAPPING AN AREA—FILLING IN THE DETAILS

You might be interested in mapping more than just a boundary line. Depending on the purpose of the traverse, you may want to fill in additional information along the way. Here are four techniques to record these details:

❖ *Sketch in features.* As you walk, sketch in pertinent features that you can see to the sides of the traverse line, estimating distances and angles.

❖ *Take sideshots to specific features.* As the traverse proceeds, the leader watches carefully to see if there are any important features near the traverse line that the team would like to show on the map. These features may be boundary markings, sacred sites, roads, landmarks, or whatever. When the leader sees one, he or she establishes a station at this point along the main traverse.

The team then takes a *sideshot* by taking a bearing from the traverse line to the feature and uses the tape to measure the distance. The procedure is the same as for measuring between a pair of stations on the main traverse. Write in your notes the distance from the last station at which the sideshot is taken, the bearing to the object, the length of the sideshot, and the name of the object to which it was taken. Again, a bearing at right angles will generally be most accurate, unless there is an angled road or trail directly to the feature. (The perpendicular distance from the traverse line to a parallel line or to a point can be called an *offset*.)

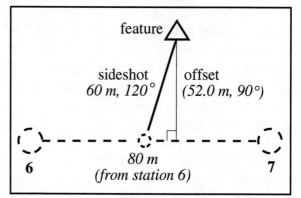

To locate a feature precisely, record both the distance to the feature and the bearing from a known point on the traverse line.

❖ *Take sideshots in a radiation pattern.* Stand at each station and take sideshots to several features. A systematic way of doing so is to check to the north, south, east and west and note what you see in each of the four directions.

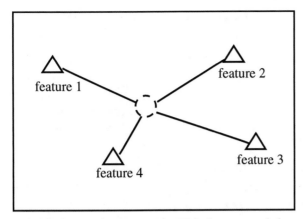

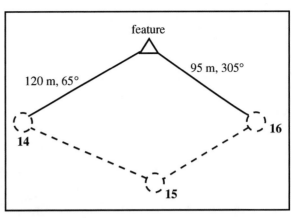

Sideshots in a radiation pattern. Take bearings to help draw the features at the correct angles. Estimate or measure the distances.

If the feature is too far away to conveniently measure the distance, then take bearings to the feature from two stations and plot the location using intersection.

❖ **Use the intersection technique.** (Note: intersection is described in subsection 8.3.4.) For example, take a bearing to a feature from station 14. Then from station 16 take another bearing to the same feature. Plot the bearings on the map; the feature is located where the lines cross.

If the primary purpose of the current traverse is to create a framework for a more detailed map, you may want to begin by traversing the outer boundary and just recording only the survey of the line itself first if you do not have time to fill in all the detail as you go. Later you may choose to come back to the traverse line to start another traverse line, or to fill in detail, or to check a mistake. To allow you to find your location on the traverse line, the stations have to be marked in some way. Take good notes while doing the original traverse. Wherever there is a large tree, a creek crossing, a road entering, a gravestone, or anything distinctive, make a note of it—especially if it's at a station. Select stations at distinctive natural markings where practical, but without significantly changing the line of the traverse.

9.4.3 RECORDING THE TRAVERSE

For ease of use, prepare a notebook or forms before you begin surveying. If you are using a notebook, write at the top of the first page the details of where the traverse is, and the date. If you are using separate forms, be sure to label each form so that you know which survey it is part of and where it belongs in the sequence, and file them in an orderly way, in binders or file folders.

In the field, record the measurements and notes in the columns, and use the prepared space to draw a sketch as you go. The sketch may be a transect sketch (subsection 5.3.1) or a panorama sketch (subsection 5.3.3).

Always start at the bottom of the page and work toward the top when writing the measurements and making notes and sketches. Why? Just imagine that you are walking forward with the page in front of you—it is easier to draw from the bottom of the page to the top, because the features on the land line up with your drawing.

It is important to clearly describe the exact starting point of the survey, so that you or someone else can find it again. Use references to fixed and identifiable features such as boulders, river junctions, or road junctions. Write that information in the first line at the bottom of the page. On the second line up, write the front and back bearings and distance between station 0 (POC) and station 1. On the third line up, write '1' in the

SAMPLE PAGES FROM A SURVEY NOTEBOOK

STN. #	FRNT. BRG	BACK BRG	DIST. (m)	SLP. (°)	NOTES LEFT	RIGHT
10						
	28	208	18	10		
9					House—Mr. T	
	6	186	13	30		
8					Creek flows to SW	
	6	186	13	30		
7					Farm hut	
	343	164	22	10		Coffee
6						82° to church
	311	131	26	0		
5					Sacred tree	348° to church
	248	68	40	5		
4					Garden hut	pond
	134	314	19	20		
3					Creek crosses, flows to NW	
	134	314	17	5		
2					44° to Mt. B	299° to Mt. A
	168	348	30	0		Rubber plantation
1					112° to Mt. B	230° to Mt. A
	122	302	20	0		
0	POC at north end of highway bridge crossing the Paku River					

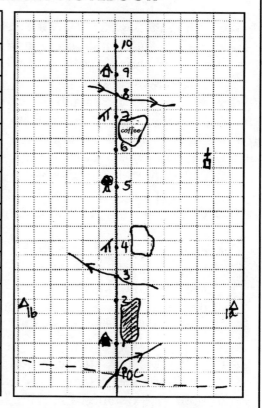

This format is standard for recording mapping notes in the field. Write your notes on the left-hand page and sketch the survey on the right-hand page as you go. Develop your own symbols for the field sketch. Make the field sketch in a straight line. Later, when you plot the bearings and distances (subsection 9.5.1), you will draw the actual route of the survey.

'station' column. On the fourth line up, write the bearings and distance between station 1 and station 2. Fill in the column for horizontal distance later after making the calculations (see subsection 9.3.3 and appendix B). Feel free to write notes at each station about features that you see there, such as large trees, special landmarks, or land ownership.

If you take any sideshots, for each sideshot use a separate line, if necessary, to record the bearings, the slope, and the length of the sideshot from the traverse line. The feature must be important if you made a sideshot to it, so be sure to write notes about it.

On the right-hand side, draw a sketch as you go. Draw the POC at the bottom of the page, draw a line up the page to station 1, etc. Draw the line straight regardless of whether your route turns. Estimate the scale, perhaps using 1 cm for every 20 m. Mark every station so that you can remember where features are in relation to the stations. Draw pictures and make notes of the things that you see along the way. This drawing is just a rough sketch that you can refer to when you plot the map. For accuracy you should refer to the bearings and distances that you've recorded in the columns.

When the field notes are complete, they describe a traverse line and all the extra details that you may want to draw on a map.

9.5 PLOTTING A TRAVERSE FROM FIELD NOTES

9.5.1 STEPS TO PLOT A TRAVERSE

Note: The following instructions assume that 'north' means 'true north' and that, if necessary, declination has been accounted for in all bearings.

Plotting a traverse is the process of using the columns of numbers in your notebook to draw a map. You need graph paper, a pencil, a ruler, and a protractor. Plot the *frame* of the traverse (That is, only the stations and the lines that link them) before adding the sideshots and the detail from your 'extra notes' column. Here are eight steps for plotting the traverse:

1. **Cut a sheet of graph paper to size.** Make it the size that you would like the map to be.

2. **Check over the field data.** Look for possible errors. If there were two people taking notes, check that they do not disagree. Check the front bearings and back bearings to make sure that they are 180° different. If they are not, either decide which one is right or split the difference. You might want to write out the data again on a clean sheet of paper.

3. **Convert slope distance to horizontal distance.** Make an extra column, if necessary, for the horizontal distance. Calculate the horizontal distances as explained above in subsection 9.3.3.

Plotting a traverse.

4. **Choose the map scale.** First, estimate how many metres long the traverse is. (For an open traverse of a route that is more or less in one direction, simply add up all the distances recorded between stations to get the approximate distance that you want to map. For a simple closed traverse, look in the data for the longest side). Convert the distance from metres to centimetres and divide the length of the traverse by the length of the graph paper.

> **scale = length of traverse (cm) / length of paper (cm)**

For example, if the paper is 90 cm long, and the traverse is 200 m or 20,000 cm long, then the scale is 20,000 divided by 90, which equals 222.22, and which is not very convenient to work with. Instead, round the number up to get a smaller scale, say 1:250 or 1:300.

Note: For more-complex traverses with many stations and a variety of bearing directions, you may first want to do a quick small-scale sketch to find out how the traverse will occupy the graph paper.

5. **Convert the horizontal distances to map distances.** Make another column in your notes to write the map distance. First convert each horizontal (ground) distance from metres to centimetres and then divide by the scale. (Also check appendix C.)

> **map distance (cm) = horizontal distance (cm) / scale**

6. **Draw a north arrow on the graph paper.** Draw the north arrow so that it indicates north at the top.

7. **Prepare to plot the bearings of the traverse.** The easiest kind of protractor to use for plotting bearings is one that is completely round or square, a full 360°. First decide where to start on the paper, so that the plotting won't go off the page. To do so, begin by placing the protractor somewhere on the paper so that 0°/360° is at the top. Look at the bearings from the first several stations and then look at the protractor to get an idea in which direction the lines will be drawn on the paper, given that north is at the top. For example, if you are plotting an open traverse that has most of the bearings at the beginning between 90° and 180°, and then they switch to a range of 225° to 360°, then you know that the survey is going to run clockwise on the paper, and that you want to start at the top of the paper, perhaps near the centre of that edge. Mark your starting point (POC or station 0).

8. **Plot the bearings.** Place the protractor so that its exact centre is at the POC. Be sure to align it on the graph paper so that north is facing the top of the page, and the line between 0°/360° and 180° on the protractor is exactly aligned with the lines of the paper. Read from your notes the bearing from the POC to station 1. Find the bearing, or angle, on the protractor, and use your pencil to make a small tick on the paper at that angle. Move the protractor out of the way. Read the map distance (calculated in step 5, using the slope-corrected distance) for this bearing from the 'map distance' column. Use the ruler to measure off this distance from the POC along a bearing through the

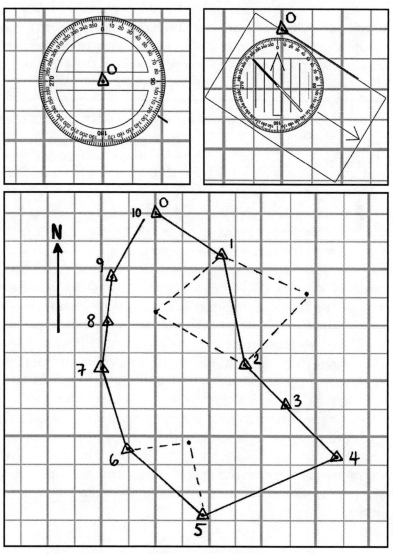

Plotting traverse data onto graph paper. You could use a compass instead of a protractor. (Data is from survey notebook pages in subsection 9.4.3.)

tick you made. At the distance you measured, make a mark to indicate station 1. As an example, if you've recorded 50 m between the POC and station 1 in your notes, and the scale is 1:1000, then the line that you'll draw on the paper will be 5 cm long. Now place the protractor, again aligned to north, so that its centre is at station 1 and then mark and measure the bearing to station 2. Continue in the same manner for all of the stations.

Note: If the line runs off the paper at any point, then carefully align additional graph paper with the piece that you are drawing on and glue or tape it into place.

9.5.2 CALCULATING THE ERROR OF A CLOSED TRAVERSE

If you did a closed traverse (in which you start and finish at the same place), when you've finished plotting the map, the lines should meet at the ends. Often, because of small errors in measuring angles, distances, or slopes, the lines do not meet exactly. Don't worry. This *error of closure* happens to professionals too, but it should be small. There is a simple way of calculating whether the traverse is OK or whether the error is too big to ignore: Measure (in centimetres) the gap between the starting point and the end of the line. Add up all the slope-corrected distances in your notes to find the *perimeter* of the traverse, and convert the perimeter to centimetres according to the mapping scale that you used. Divide the gap by the perimeter, multiply by 100, and you have *percent error*. As a general rule, if the error is below 3%, you have a good traverse.

---

percent error for a closed traverse = gap (cm) / perimeter of traverse (cm) × 100%

Example:

2 cm / 218 cm × 100% = 0.92% error

If the error is larger than 3, then look for the sources of error and try to correct them. Go back through your notes and look at the plotted map to see if there are any distances or bearings that don't make sense. If you've made an error writing down the numbers, sometimes it can be seen in the shape of the traverse. If you took back bearings, check them—they should be 180° different from your front bearings in each case. If you find an exception, you may have solved the puzzle. But often there are several small cumulative errors that are more difficult to locate.

Maybe you held the compass wrong and made consistent mistakes taking bearings. (Note that if the declination was set wrong on the compass, the plot will still be the right shape, but it will be rotated around the POC, and you probably won't notice anything wrong until either you use a different compass or a GPS receiver, or you compare the map with another one for the area.) Maybe the metre tape was sometimes appropriately tight and sometimes loose, or maybe at times it zigged and zagged around obstacles instead of going straight. Perhaps you or the other person held the metre tape in the wrong place or you read it wrong or made a recording error. Maybe you forgot to account for slope on one or two readings or did so incorrectly, or maybe you measured wrong while plotting.

9.5.3 DESIGN AND DRAW THE MAP

Up to this point you've only drawn the stations and the connecting lines on the graph paper. Once you are satisfied with this frame, you want to draw a map. Here is a brief description of how to continue—you can read more in chapter 12.

Cut a sheet of tracing paper large enough to cover the plotted traverse, leaving space for a legend on the side (or bottom). Draw a frame for the map and align the frame with the north lines of the graph paper. Draw a frame for the legend.

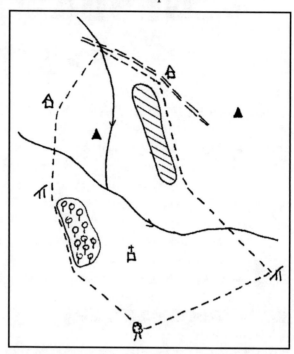

If you want the actual line of the traverse on the map (for instance, if it follows a boundary or a trail), then trace the traverse line on the tracing paper. Don't draw the stations themselves or write the station numbers.

If the ends of a closed traverse don't quite meet and the gap is small, it's okay to draw a short line to close the traverse so as to result in a neater looking map. However, a more accurate way of depicting the survey is to distribute the error over several legs of the traverse. Be sure to leave the gap on the original plot on the graph paper to show the actual result of the survey.

Lay tracing paper over the traverse that you plotted on graph paper and draw the information that you want to show on the map. The use of symbols is an important part of producing a map from survey data.

If the traverse line itself is not important, then simply use the plotted traverse for reference to locate the features on either side of it.

Review your notes and design the symbols for all the features that you would like to draw on the map. Then decide how to arrange the legend and other necessary information, including

❖ *Title*
❖ *Descriptions of each of the symbols used*
❖ *Scale*
❖ *North arrow*
❖ *Date surveyed and date drawn*
❖ *Who surveyed and who drew*

Draw the features—such as houses, gardens, small rivers—from your sketches and notes by locating them in relation to the stations. Read subsection 4.2.5 and section 12.2 for more ideas about designing symbols and map format.

10 SURVEYING WITH THE GLOBAL POSITIONING SYSTEM (GPS)

What you will learn in this section:
- ➤ What the GPS is
- ➤ How the GPS works
- ➤ How to set up a GPS receiver for your area
- ➤ How to assess the accuracy of GPS coordinates
- ➤ How to use the GPS in the field to find the coordinates of locations
- ➤ How to record GPS data
- ➤ How to find a location on a topographic map by using its GPS coordinates
- ➤ How to make a base map using GPS data

10.1 ABOUT THE GPS

10.1.1 WHAT IS THE GLOBAL POSITIONING SYSTEM?

The Global Positioning System (GPS) is a navigation system that consists of satellites in the sky and, here on Earth, instruments that receive signals from those satellites. Twenty-four special GPS satellites are constantly orbiting the Earth. A hand-held GPS receiver receives signals (like a radio) from the satellites and uses information from these signals to compute the near-exact location of the receiver on the surface of the Earth.

By using one of these GPS receivers, you can learn the near-exact location of any point on the land—a sacred site, a boundary landmark, the corner of a garden, a house—in terms of a standard coordinate system. Remember that coordinates are given with a pair of numbers that refer to a specific place on the Earth. Many maps, such as topographic maps, show a coordinate system (be sure to check the legend to see which one) with lines and numbers along the sides of the map (*latitude* or *northing*), and along the top and bottom (*longitude* or *easting*). You can find

The Global Positioning System is composed of 24 satellites in orbit around the Earth, control stations belonging to the US military (USDoD), and hand-held receivers that any of us can buy.

the location on the map that corresponds to a particular pair of coordinates by finding the two numbers along the edges of the map and using a straightedge to determine where perpendicular lines through them would cross. (The process is described in detail in subsection 10.4.1.) You can do this procedure in reverse if you want to read off the coordinates for a particular place on the map. (You can review coordinate systems in subsection 2.2.4.)

A GPS receiver uses sophisticated technology but is simple to use.

When you are standing on the ground, it is often difficult to know exactly where you are on the map unless you are near a recognizable landmark. The GPS can tell you where you are on the map, even if you are standing in the middle of a swamp or on the side of a mountain far from a definite landmark.

Although the GPS employs very sophisticated technology, it is very simple to use: just turn the GPS receiver on and wait for it to display the coordinates of your location. (The more difficult part is knowing how to use the coordinate information to make maps.) The only reason you need to know how the GPS works is so that you can estimate the accuracy of the coordinates that the receiver gives you, and to troubleshoot the situation if something seems to be wrong.

10.1.2   HOW DOES THE GPS WORK?

The basic principle of the GPS is quite easy to understand. Remember how (in subsection 8.3.4) we talked about triangulation with a compass? If we know the directions to three locations, we can triangulate to find the location where we are. Well, a GPS receiver works by measuring the distance to three (or more) satellites that are within its 'field of view.' The receiver knows where each of these satellites should be at any given time, because it has an almanac (like a bus schedule) stored in its memory.

But how does a GPS receiver measure the distance to something so far away? It does so by timing how long the signal takes to arrive from the satellite and then calculating the distance based on the speed at which radio signals travel. But how does the GPS know when the signal left the satellite? The signal is coded to indicate when it left the satellite. The GPS receiver reads the code and then calculates the difference between the signal's departure and arrival times. Because the distance is calculated based on precise timing, the satellites have been designed with extremely accurate (and expensive) clocks. The receivers have less sophisticated clocks in them. Errors caused by the receiver being slightly out of synchronization with the satellites can be compensated for by calculating the distance to a fourth satellite.

10.1.3   FACTORS AFFECTING THE ACCURACY OF A GPS RECEIVER

There are several factors that affect the accuracy of every GPS calculation of position.

- ❖ *satellite clock error:* Satellites have highly accurate atomic clocks, but there is always a small margin of error.
- ❖ *ephemeris error:* Each satellite's position can vary from the calculated orbit because of gravitational pull from the sun and moon. The satellites are monitored at the control stations by the US military (USDoD) and their positions are usually corrected.
- ❖ *receiver error:* The clock in a receiver always has more error compared to the accuracy of the satellite clocks. This error is largely, but not completely, compensated for by triangulating from four satellites instead of just three.
- ❖ *atmospheric disturbance:* Constant variations in the ionospheric layer of the Earth's atmosphere speed up or slow down the signals, thereby making the distance calculations slightly incorrect.
- ❖ *'selective availability':* The USDoD, who developed the system, intentionally and intermittently interfere with the satellite signals so that we never know when our GPS receiver is giving a truly accurate position and when it is not. This tampering is called 'selective availability' (s/a). Its purpose is to prevent the general public and foreign militaries from using the GPS to get super-accurate locations. This source of error is both the most significant one and the one that we can do the least to predict or control. If selective availability is ever discontinued, it will greatly improve the accuracy of non-differential GPS measurements (see subsection 10.1.4), by 2.2 times, making them much more useful for map-making, but still not as accurate as differential GPS coordinates.

ESTIMATING GPS ERROR

ERROR SOURCE	POSSIBLE ERROR* (metres)
Satellite clock error	0.6
Ephemeris error	+0.6
Receiver error	+1.2
Atmospheric (ionospheric) error	+3.6
Subtotal without s/a (rounded up from 3.9 for convenience)	**4**
Selective availability (s/a) if implemented	+7.5
Total with s/a (rounded up from 8.5 for convenience)	**10**

To estimate the accuracy of your GPS coordinate, multiply the total above by the PDOP (Precision Dilution of Position, described in subsection 10.2.2) for that coordinate.

<u>Example:</u> If the PDOP is 3.5, then the accuracy of your GPS coordinate is

With s/a implemented (3.5 × 10) = ±35 m

Without s/a implemented (3.5 × 4) = ±14 m

*Note: Total error = sqrt ((error 1)^2 + (error 2)^2 + ... + (error n)^2)

—*from Trimble, 1989*

A further source of error, one that we can work around to some extent, is a result of the configuration of the satellites. Remember that triangulation is most accurate if the points that we are triangulating from are at a wide angle from each other relative to where we are standing. If the satellites are all clustered in one place in the sky, the calculation of the position will not be as accurate as if they were distributed widely. Each of the 24 GPS satellites moves in its own orbit or path around the Earth, so they are always changing configurations in the sky. When the satellites are configured so that several are 'in view' of the receiver and spread out across the sky, then the calculated location will be the most accurate.

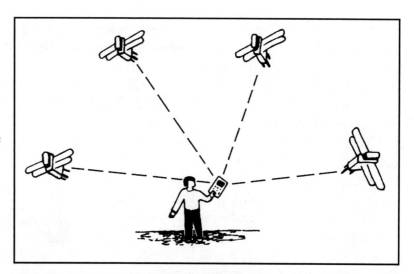

Good (top) and bad (bottom) satellite configurations.

By the way, for a satellite to be 'in view' of the receiver does not mean that you need to see it—it means that the receiver can get a signal from the satellite. Usually that means that you need to have a good, wide view of the sky unobstructed by mountains, buildings or even tall trees. At the bottom of a valley, for instance, fewer satellites are 'in view' and so there is less chance that the few satellite signals that the receiver does pick up are in a good configuration. In valleys you might have to wait several hours until the satellites move into a good configuration.

10.1.4 TYPES OF GPS RECEIVERS

There are many makes and models of GPS receivers for sale. They come in different shapes and sizes, with varying quality and price. They all work on the same principles, however. There are two main categories of GPS receivers: *differential* (survey-quality) and *non-differential* (recreational-quality). The main difference between these categories is how they compensate for error.

Differential GPS: A differential GPS system consists of at least two receivers. One is set up as a reference receiver. It calculates correction data and sends it via a radio signal to remote receivers, which use the data to correct their coordinate positions to an accuracy of within 2 m (read about differential GPS in subsection 10.2.4). Differential GPS receivers are expensive, typically US$3000–10,000 for each receiver.

Non-differential GPS: Only one receiver is needed. It is simpler to use than differential GPS and cheaper, typically US$150–800 for one receiver. Measurements are usually accurate to within 30 m; at worst, to within 100 m. Some non-differential GPS units can compensate for error by averaging. A receiver with averaging capability can be placed in one location and set up to calculate its coordinates every few minutes over several hours and then provide a more accurate average value (read about averaging in subsection 10.2.4).

In both categories of GPS, the quality of both the antennae and the tuning mechanism have improved greatly in recent years. Newer models have 12 channels, giving them a greater capacity to receive several satellite signals at one time. This feature means that we can get a 'position fix' faster, even under a thin forest canopy or at the bottom of a valley.

This unit is just one of many different models of GPS receiver that all work on the same principle and with the same satellites.

10.2 HOW TO FIND THE COORDINATES OF A LOCATION USING THE GPS

First 'set up' the GPS receiver for the local area, as described in the next subsection. Then carry it to the first place for which you want to know the coordinates. Stand in a clear area with no obstructions overhead. Turn the receiver on and wait until the coordinates (possibly called 'waypoint') are displayed. Write the coordinates in your notebook or on your GPS field data form (see appendix A). It is as simple as that.

10.2.1 ADJUSTING A NEW GPS RECEIVER FOR THE LOCAL AREA

The first time that you use the GPS receiver in any local area, it must be adjusted for that region on Earth. The receiver stores and automatically updates the positions of all the GPS satellites in an almanac. The almanac memory area is empty when the receiver is first removed from its box, or if the batteries run out and the memory is lost. To collect the current almanac data, take the receiver outdoors and place it with the antenna facing upward with a clear view of the sky. Leave it that way for about 15 minutes.

When first using the GPS receiver in a particular local area, there are also adjustments that you need to make manually, by pushing buttons. Each of the various models of GPS receiver has different buttons to make the adjustments. Refer to the manual for your GPS receiver to know how to do the setup adjustments. Here are the basic factors that are critical to adjust in most models:

❖ **Time:** The clock has to be set for the time zone of the local area. The time zone is given as the number of hours ahead of (+) or after (–) Greenwich Mean Time (GMT), which is also known as Universal Time (UT). The instruction booklet will give you a chart to know the time zone for the local area. Set the time zone before collecting an almanac.

❖ **Coordinate system:** Set the receiver to display either **UTM** or **latitude and longitude**, whichever grid is on the topographic map that you are using. If the map has both, then it may be easiest to use UTM, because it involves simpler measurements (and it can also be more accurate). If you don't have a topographic map, then use **latitude and longitude**.

❖ **Distance and elevation units:** Choose **metres**.

❖ **Datum:** A datum is a cartographic correction system (using a particular *reference ellipsoid*) that compensates for irregularities in the roundness of the Earth. Every topographic map series is made using a specific datum. A GPS receiver will have 25–100 different choices of datum, but the receiver will reference all of them to the WGS84 datum (where 'WGS' stands for 'World Geodetic System'). A topographic map should have the title of a specific datum written on it; for greatest accuracy, use that datum if your GPS has it. If not, call the local government survey department and ask what datum they use. Most receivers have an option to custom program a datum. To do so, you need to enter five long numbers. To find out what numbers to use, you could again ask the government survey department. If you are unable to find a specific local datum to use, use WGS84, which is the most common datum. Your GPS should give you accurate enough coordinates if you use this one, particularly if you are making a map from scratch.

10.2.2 READING THE GPS RECEIVER

Each model of GPS receiver has a different configuration of display and different buttons to press to show the information that you want. You will have to practice with your own receiver to figure out the buttons. Most of the time, the only information that you will need is the coordinates, which are shown automatically, and a number that relates to the accuracy of those coordinates, usually the *Precision Dilution of Position (PDOP)*.

The PDOP is based on the distribution of the satellites that the unit received signals from. When the receiver has calculated its position from at least four satellites, it calculates an estimate of the accuracy according to the 'geometry' or distribution of the satellites at the time of the reading. Any PDOP number under 7 is usable, but the smaller the better. We usually recommend waiting until you have a PDOP of less than 3. Why will waiting help? Because the satellites are always moving relative to each other and to you, if they are clustered together now (giving a high PDOP), they will eventually move into a better configuration (giving a low PDOP). (To better understand the relationship between PDOP and accuracy, review subsection 10.1.3.)

Some GPS receivers give an *Estimated Position Error (EPE)*, which is automatically calculated from the PDOP to give an estimate of the error directly in metres.

Read the coordinate and record it in a notebook.

The GPS receiver also displays information about the status of each of the 24 GPS satellites. Every receiver model displays this information differently: some do it graphically, and some use special terminology. In general, the information that you will find interesting is

❖ *satellite number*
❖ *status of each satellite; for example whether the receiver is **searching (s)** for the satellite, whether the receiver is **locked (l)** onto the satellite, or whether the satellite is 'unhealthy' (h)*
❖ *signal strength for each satellite in view; shown by a bar graph*
❖ *position of each satellite by azimuth (bearing from north) and the angle above the horizon, both shown in degrees*

This data is for your own interest only, to help you to use the GPS more effectively, and need not be recorded. For example, if you don't understand why the receiver is taking so long to calculate coordinates, look at the satellite position and status and you might find that the configuration is poor because not only are there just a few satellites in view, but one or two have poor signals.

To use a GPS receiver effectively, choose carefully where you use it. The GPS unit can receive satellite signals only if there is no obstruction above the antenna (which is usually built into the receiver). This limitation means that a GPS unit is useless in dense forest, in very steep valleys, or inside buildings. For some receivers you can buy an external antenna that you can mount on a pole to reach above a low forest canopy. If you need to know the location of a special site in the forest, you may be able to get the receiver to work if you cut some tree branches to get a clear view of the sky. However, it is usually less work and more effective to take a GPS reading from a nearby clearing. Then measure over to the site in the forest with compass and metre tape.

10.2.3   RECORDING GPS DATA IN THE FIELD

Most GPS models allow you to record the position data in the receiver with a push of a button. But it only records the coordinate number and a short name that you may give it. Generally, you will want to record a lot of information about a place while you are there, so it is best to also write it down on a form or in a notebook. Another reason to write it down is because there is always a risk of losing the position data that is stored in the receiver—if the user makes a mistake or if there is an extended power loss.

If you have walked eight hours to a landmark on the boundary of the village's land, you want to make sure to take the time to get a good 'fix,' or coordinate, and that you write down all the important information. You don't want to have to walk back to the landmark a second time because you didn't get enough data or you are not sure about the quality of it, or because you lost it. You can make up special forms for recording GPS coordinates (see sample in appendix A) or you can make a special page in your usual field notebook. You should record this information:

❖ *Waypoint number:* Each place that you take a GPS reading will have its own number. Later, these numbers will identify the location on a map, so it is important that there is no confusion. Often the numbers are written like this: W001, W002, W003, W004. If you are working with more than one GPS receiver at a time, then you should work out a system to assign numbers so that each location has a unique number. For instance, one team could label its waypoints WA001, etc. and the next team could use WB001, etc. and so on.

- ❖ *Name of the location:* Use the local language name and be as precise as possible. Give the name of the rock or the river mouth or the headwaters. The purpose of naming the place precisely is to identify it clearly to help in finding it again if you want to, or so that you can ask the elders about it.
- ❖ *Coordinates:* Write down all the digits in each number, whether in latitude and longitude or in UTM.
- ❖ *PDOP or EPE number:* Write down the PDOP or EPE number. (Note that if you used averaging there will not be a PDOP number.)
- ❖ *Averaged position or differentially corrected position:* Did you use averaging or differential correction? (See subsection 10.2.4.)
- ❖ *Notes:* Describe the location. *What type of land use? Who owns it?* Take compass bearings and note what you see to the north, south, east and west.
- ❖ *Date:* Record when you collected the GPS data.
- ❖ *Surveyor:* Record the surveyor's initials for every position. Later, if the person who is drawing the map has any questions about the data or about the place, then he or she knows whom to ask.

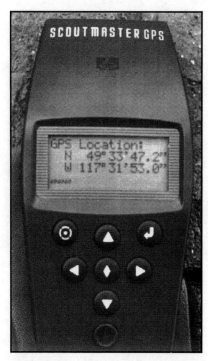

An example of how a GPS receiver displays the coordinates of a location. Other models will look different.

10.2.4 IMPROVING THE ACCURACY OF GPS DATA

GPS receivers use timing signals from at least four satellites to establish a position. Each of those timing signals has some constant error or delay that can depend on many factors, as described in subsection 10.1.3. But recall that 'selective availability' is the most significant cause of error. There are two common ways of improving or correcting the data: averaging readings over time and using a differential receiver system.

Averaging

Averaging is done by a single receiver, using a special averaging function. The receiver is left open and on in one place for a period of time—from one to seven hours—while it calculates the position repeatedly and calculates the average of all the coordinate readings. If there is error on account of selective availability or atmospheric conditions, the error can be 'averaged out.' Averaging needs to be done for a minimum of one hour in order to be useful. The longer the period of averaging, the better the reading. However, after about seven hours the accuracy increases very little, because selective availability is usually in effect only for very short periods of time.

Differential GPS

Differential GPS requires two GPS receivers. One is the *reference receiver* (or *control receiver* or *base station*), which is set up at a location for which the coordinates are known exactly (within 1 m). The *mobile receiver* (or *field receiver* or *remote receiver*) is carried around to locations for which we want accurate coordinates. Both receivers receive the same GPS signals at their different locations. But the reference receiver also calculates backward, using the known position to calculate what the travel time for each GPS signal *should* be. Then it calculates the timing error, or **correction factor**, by comparing the actual timing with what the tim-

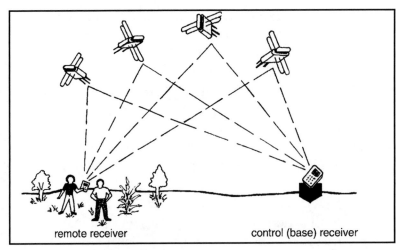

The control receiver, at a fixed, known location, receives signals from the same satellites as the remote receiver and calculates correction data for it.

ing should be. The reference receiver doesn't know which satellites the mobile receiver is actually using, so it calculates correction factors for all the satellites within its view. The mobile receiver gets the complete list of correction factors (by immediate radio transmission or later—the two systems are described below) and applies the corrections for the particular satellites that it is using (or used).

In summary, for differential GPS you need two receivers, one that can be set up as a reference receiver to calculate backward and provide correction factors and another that is set up to receive the error corrections and incorporate them in the coordinate calculations. However, GPS receivers with this capacity are usually about ten times (or more) the cost of ones that don't (that is, about US$3000–10,000).

Corrections (calculated by the reference receiver) for a given satellite are applied to the field data (by the mobile receiver) for the same satellite and time. The more satellites that are shared by the two receivers, the more corrections are available and the better the final position accuracy is. To ensure that both receivers are using the same satellites, the remote receiver should be within 100 km of the reference receiver. It is best to set up the base receiver in a place where it has a wide field of view, so that it has the potential to receive signals from as many satellites as possible.

Described below are three ways of processing the differential correction data. **Real-time** and **post-processing** are equally accurate, that is to within a few centimetres. The difference is that real-time processing is achieved faster. **Manual processing** is slower to do, and much less accurate.

Real-time Differential

With real-time differential correction, the reference receiver transmits corrections directly to the mobile receiver via radio. Therefore, the corrections are applied by the mobile receiver at the same time that it calculates its position coordinate. Real-time differential is necessary when you need highly accurate positions very quickly. To do real-time differential, you need two GPS receivers with differential capacity and some kind of communication link, such as radio. In some places you may be able to get differential correction data from already established (often government owned) reference stations that broadcast the differential correction information, usually on longwave in the 300 kHz range.

Post-processing Differential

With post-processing differential, there are also two receivers receiving satellite signals and calculating positions, but the data is stored in both receivers, rather than being transferred directly from one to the other. After the survey, the data is downloaded from both units into a computer, the data is corrected, and the corrected coordinates are printed out. To do post-processing differential GPS you need two GPS receivers with differential capacity, computer software (which is sold with the receivers), and cables to connect the GPS receivers with the computer. For most models, the base unit and remote unit are identical.

Manual Differential

If you don't have true-differential GPS equipment, there is a non-automatic method for doing differential corrections. You still need two GPS receivers—the same model if possible—one to keep at a base (or 'reference' or 'control') station and one to use at remote (or 'field' or 'mobile') stations. The most difficult part of this method is determining a reference point. It is best to use a government-surveyed triangulation point.

MANUAL DIFFERENTIAL GPS—AN EXAMPLE

The actual coordinates of a reference or control station must be known in order to use manual differential error correction. For instance, the station might be at a government-surveyed triangulation point. Simultaneously record GPS readings every 5 minutes for one hour at both the reference and the remote locations. Because the location of the reference station is known, you can calculate correction factors for each reading at the reference station: **correction factor = actual coordinate – reference GPS coordinate***. Then apply these correction factors to the readings from the remote station:* **corrected coordinate = remote GPS coordinate + correction factor***. Average the corrected coordinates. (This example—from Momberg, Sirait, Atok, 1996—averages just four readings, which will give less accurate results.)*

REFERENCE STATION

TIME	EASTING			NORTHING		
	ACTUAL COORDINATE	REF. GPS COORDINATE	CORRECTION FACTOR	ACTUAL COORDINATE	REF. GPS COORDINATE	CORRECTION FACTOR
10:00	534,654	534,664	–10	9,725,010	9,725,014	–4
10:05	"	534,683	–29	"	9,724,982	28
10:10	"	534,632	22	"	9,724,971	39
10:15	"	534,686	–32	"	9,725,069	–59

REMOTE STATION

TIME	EASTING			NORTHING		
	REMOTE GPS COORDINATE	CORRECTION FACTOR	CORRECTED COORDINATE	REMOTE GPS COORDINATE	CORRECTION FACTOR	CORRECTED COORDINATE
10:00	533,231	–10	533,221	9,726,414	–4	9,726,410
10:05	533,215	–29	533,186	9,726,462	28	9,726,490
10:10	533,242	22	533,264	9,726,453	39	9,726,492
10:15	533,197	–32	533,165	9,726,376	–59	9,726,317
Average of corrected coordinates:			533,209			9,726,427.25

The two receivers must be using the same satellites. If the receivers are older models (before about 1996) that use only a few channels, then you must be able to set up each receiver to select the satellites that it uses in its calculations—through radio contact, the operators can agree to use the best four satellite signals. The newer twelve-channel GPS receivers (if they are not more than 100 km apart) are very likely to be receiving signals from the same satellites, and therefore selecting the satellites is not as critical.

The operators at both the base and remote locations synchronize their watches (or use the clocks on their receivers) and agree beforehand on the exact times to record coordinate readings. They then take a reading every five minutes over a duration of one hour. Afterward, they compare the coordinates reported by the control or reference receiver with the known coordinates for the base location and calculate the difference for both the longitude and latitude (or the easting and northing) coordinate taken after each five-minute time interval. The result will be a list of numbers, or *correction factors*, one for each reading, to apply to the data from the remote receiver. The first four entries for the correction data are shown in the table on the preceding page.

To apply this correction data, add to or subtract from the data recorded from the remote receiver. In the example in the sidebar, at 10:05 the correction factors are –29E and +28N. The reading on the remote GPS receiver at 10:05 is 533,215E and 9,726,426N. Apply the correction factors by subtracting 29E and adding 28N. Therefore, at 10:05 the remote GPS has corrected coordinates of 533,186E and 9,726,490N. After correcting the coordinates for each time interval, average all the corrected coordinates to improve the accuracy.

The survey team will need to spend at least one hour at every location for which they want to do this manual differential process (and at least one additional person will be needed to record the coordinates at the control station). As with averaging, which takes up to seven hours, it is best to select priority places for which you want to take this much time to get accurate coordinates.

10.3   STEPS TO PLAN A GPS SURVEY

Communities can make maps very quickly and efficiently using the GPS. Most communities, however, cannot afford the highly accurate differential GPS receivers. If you are in this situation, you need to plan how to use non-differential GPS receivers appropriately. The following nine steps will help you to plan a GPS survey, assuming that you do not have a *differential* GPS receiver setup (which is expensive).

1. *Decide whether or not to use the GPS,* according to the purpose of the survey and the size of the area (see sidebar on the following page).
2. *Determine the datum* to use, according to the topographic map.
3. *Determine the type of grid* that you want to use—UTM or latitude/longitude.
4. *Determine the priority place(s)* for which you want to know highly accurate coordinates, and for which you are prepared to use *averaging* or *differential* (manual) methods. Examples include the centre of the village, at the start of a compass survey, or at important landmarks on the boundary.
5. *Determine the types of other places* for which you want to know the coordinates with less accuracy—for example, landmarks on the boundary, sacred sites, historical sites, or sites with significant economic use. If you have a topographic map, the highest priorities are features that are difficult to locate on the map (for example, located in a flat forest). The lowest priorities are places that are easily located on the map (such as mountain peaks and river junctions). But, if

WHAT TO USE A NON-DIFFERENTIAL GPS RECEIVER FOR

✔ *Mapping large areas at a scale of 1:10,000 or smaller*

✔ *Mapping large areas for which you have no topographic map*

✔ *Mapping large areas that are flat or uniformly hilly and therefore difficult to interpret on a topographic map*

✔ *Finding the coordinates of the starting point of a compass survey*

AND WHAT *NOT* TO USE IT FOR

✘ **Small areas that you need to map accurately:** *A compass traverse is much more accurate than a series of coordinates from a non-differential GPS receiver. For instance, if you want to measure and map a 20-hectare portion of the village territory that has been allotted for a plantation, it is better to use a compass traverse.*

✘ **Small areas that you want to map in detail:** *You won't save much time using the GPS because, if you are recording a lot of detail about the area as you survey, then the location measurements take only a fraction of the time that you will need for the survey. It is better to use a compass traverse, which is more accurate.*

you don't have an accurate topographic map, then prominent landmarks such as mountain peaks, rivers, and road junctions are a high priority.

6. *Divide into teams,* according to the number of GPS receivers that you have.

7. *Decide on a numbering system for waypoints* so that each team numbers their set of waypoints differently and systematically, so that when you compile the data there is no confusion between the sets of data.

8. *Determine what information* the teams will note—for example, at each coordinate will you record the name of the site? Will you record the name of the family that owns the site or the field, or will you just record the type of land use or feature there? Perhaps use the four-direction method, taking bearings to the north, south, west, and east of the point and noting what you see.

9. *Determine the location of the reference receiver,* if you are doing non-automatic (manual) differential GPS, and also the area to which you will take the remote receiver. Then agree on the coordinate-recording time intervals.

10.4   MAKING A MAP WITH GPS DATA

To make a map with GPS data, you can either find the coordinates on an existing topographic map or create a completely new map. Both ways are described below. Most community maps are made the first way, using a topographic map as a base map and using the GPS to find where to mark important locations onto the map. This way you can use all the information that is already on the topographic map, such as the locations of rivers and the shapes of mountains. If there is no topographic map available, then you can make a map using just the GPS. However, the problem with this method is that you must walk (or otherwise travel) to every point whose location you want to show on the map in order to get its coordinates. However, as suggested above, you can plot some of the points using the GPS and others using compass survey techniques, as described in chapter 9.

10.4.1　FINDING GPS COORDINATES ON A TOPOGRAPHIC MAP

It is a simple procedure to locate GPS coordinates on a topographic map. Look for the numbers of latitude (or UTM northing) on the sides of the map and look for the numbers of longitude (or UTM easting) on the bottom or top of the map. (You checked the map before you started collecting GPS points and set up the GPS to collect data in either latitude/longitude or UTM.) To find a coordinate, use a ruler to divide between these numbers (as explained below) until you find the exact number as recorded with the GPS. Then use a long ruler to draw a pencil line across the map at those points, from right to left, and top to bottom. It is important that the lines be perpendicular to the coordinate grid on the map. Where the lines cross is the location of the coordinates. How you divide the space between the numbers of the edge of the map is different depending on whether you are using latitude/longitude or UTM.

Latitude and Longitude

Latitude and longitude are expressed in degrees, minutes, and seconds. There are 60 minutes in a degree, and 60 seconds in a minute.

As an example (see below), to locate longitude 115°20'50", look for the numbers across the bottom of the map and follow these steps (use a pencil for marking):

1. Find 115° and 120°. Divide the distance between them by 5 to find where each degree is (the one next to 115° is 116° and so on). In this example you need mark only 116°.

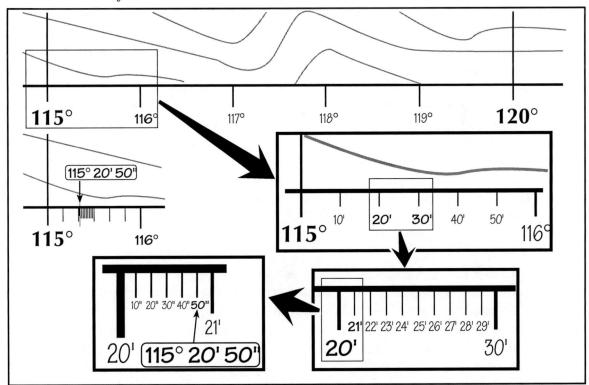

Begin by marking individual degrees. To subdivide degrees of longitude or latitude into tens of minutes, divide by 6. Then (as space allows) continue subdividing: first by 10 to get individual minutes, then by 6 to get tens of seconds, then by 10 to get individual seconds. Note: In order to illustrate the process more clearly, successive boxed enlargements are shown. (For clarity, the numbers and line thicknesses in the enlargements are not shown exactly to scale.) In real life, however, you would round this coordinate off to 115°21', because in practice you would not be able to accurately differentiate the tens of seconds on a map at this scale.

2. Divide the distance between 115° and 116° by 6 to find every 10 minutes: 10', 20', 30', and so on.
3. Further subdivide by 10 the space between 20 and 30. Make a mark at 21'.
4. Divide the distance between 20' and 21' by 6 to find every 10 seconds. Make a mark at 50". If you needed to, if the map had a large-enough scale, you could further subdivide the space between adjacent tens of seconds by 10 to mark individual minutes.
5. Use the same process of division for latitude on the side of the map.

Note: Starting with the tens gets you 'into the ballpark.' Then progress to the single minutes or seconds. You need mark only the ones that you require. If the map is not of a large-enough scale, you may be able to pinpoint only the minutes but not the tens of seconds (or perhaps just the tens of seconds but not the individual seconds).

Universal Transverse Mercator (UTM)

UTM coordinates are even easier to find than longitude and latitude, because the division is simply by hundreds or tens. That's because on a 1:50,000 scale topographic map the UTM lines are always 2 cm apart, which is 1000 m (or 1 km) on the ground.

As an example (see below), an easting coordinate number from the GPS receiver might be 0381520. Look at the bottom edge of your map until you find the numbers 0381000 and 0382000. You have now located the coordinate to within four digits. At a scale of 1:50,000, the lines for 0381000 and 0382000 are 2 cm apart. Divide this 2 cm length by 10 (into 2-mm sections) to find out where 0381100, 0381200, etc. should be, and mark off 0381500 and 0381600. You have now located the coordinate to within three digits. By measuring very carefully, you can break down the 2-mm section between 0381500 and 0381600 into a further 10 parts, each measuring 0.2 mm, and find the exact position of your coordinate located 2 of these new divisions over from 0381500, at 0381520.

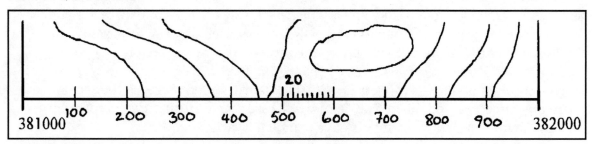

To find UTM coordinates, you need only divide intervals by 10.

You can now use the same process for the northing coordinate, using the numbers down one side of the map.

Mark the Coordinates

Mark the coordinate with a small dot exactly where the lines cross. Later you can erase the lines. Write the GPS waypoint number beside the dot.

You can often see immediately whether a GPS coordinate 'fits' the map or not. For example, you might have recorded the coordinates of a spot on top of a mountain peak, but when you plot it on the map, it looks like it is part way down the side of a mountain. It doesn't fit the map. The error could be in the GPS or in the map—you don't know.

If the coordinate location doesn't 'fit' with the map,

❖ *Check your notes again and make sure that you didn't misread the coordinates.*

❖ *Check the PDOP or EPE number to assess the quality of the data.* If it is poor, and the coordinate is for an important place, go back there and take a new reading.

You could use a marking system to clarify the 'fit' of the coordinates. If a point fits the map, mark it with a dot. If it doesn't, mark it with a dot with a circle around it. If you aren't sure, mark it with an 'x.'

Your GPS data is very valuable because it has taken you days to walk to all the places shown on this data map to collect the coordinate data. The coordinate numbers with PDOP and date are proof that you used the GPS, so store your data sheets safely. You could copy this data onto the reference map as well. Having this data on the map enhances the credibility of your work with government agencies, and it also makes it easier to use the map for reference. However, although all this data is impressive and useful for some people, it might be confusing for community members. Therefore, there is no need to write all this data on every thematic map, just the reference map. Simply list in the legend the coordinates and the names of all the places that you surveyed.

10.4.2 MAKING A MAP WITH GPS DATA AND NO TOPOGRAPHIC MAP

It is possible to make a map based on GPS coordinates alone. The principle is to create a coordinate system on blank paper and then locate the coordinates on the grid in the same way that you would find coordinates on a topographic map.

People who make maps directly from GPS data usually do it with computers. The companies that sell GPS receivers often also sell map-making software. Even the most simple GPS receivers, such as the Garmin 12, have the capacity to download coordinates into a computer to use in drawing a map. If your GPS has the capacity, you can store route data and follow rivers and roads and boundaries and have the computer automatically draw those lines. Or you can collect just point data, for example at sacred sites. This technique sounds like an easy high-tech way to make a map. It is—and, if you have differential GPS equipment, it is a highly accurate way to make a map. But remember that if you are using non-differential GPS, any one point can be up to 100 m wrong. What's more, without a topographic map you have no information such as contour lines and rivers to cross-check the data with. And, because that error can be in any direction, you might find (in a worst-case scenario) that two points that are in real life up to 100 m apart might have completely switched places!

Another problem is that a GPS map is made up only of points. If you want a linear feature on the map, for example a river, you must walk (or canoe) the length of the river with the GPS and then you (or the computer) must draw the line by connecting the GPS points. Between each pair of points the line will be straight, so the closer together that you take GPS data, the more detailed the picture, and the more detailed you can draw the bends in the river. A GPS receiver can take (for practical purposes) continuous data while you are walking along a route, but it requires a lot of memory. An inexpensive GPS receiver may not be able to store this amount of data.

It might be efficient to make a boundary map with just GPS data, but, to put the boundary in context and to fill in the map, a topographic map is very useful, if not essential.

Under what conditions would you use a non-differential GPS to make a map from scratch, given that the accuracy is so poor? Here are two of those situations:

❖ *If there are no topographic maps of the territory at a useful scale*

❖ *If the territory is too large to survey by compass, and so flat that the topographic map for it shows few or no contour lines or reference points to use in navigating or surveying (and aerial photos are not available)*

Making a GPS-based Map by Manual Means

Even if you do not have access to a computer and map-making software, you can still use the GPS data to make a map by using manual drawing techniques. The map will have all the same disadvantages and accuracy of a map made with a computer. The way to do it is to establish a grid on a large piece of graph paper, and then to mark your GPS coordinates onto this grid.

Follow the seven steps below. (This description is for coordinates in longitude and latitude. Use the same principle to make a grid for UTM data—as you can see in the table in appendix E, the measurements for UTM are simpler.)

1. ***Begin by examining your list of waypoint coordinates.*** (Although you may have surveyed many important points, it's the ones that mark the boundary of the territory that you are interested in here. Find the highest and lowest values of longitude. Round the highest value up to the nearest minute and the lowest value down to the nearest minute. Subtract the lowest from the highest. (*Note: If you are working on either side of 0° longitude, you will need to add the highest values for eastern and western coordinates.*) This number is the width of your map in minutes. Use the same process with latitude to find the height. (*If the territory crosses the equator, you will need to find the highest values north and south and add them instead.*) Now use the degrees/distance conversion table in appendix E to calculate how many centimetres wide and high the map will be at the scale that you want to use.

2. ***Choose a piece of graph paper that will be big enough for your map, including margins.*** Draw a rectangular frame on it the size that you just calculated, in centimetres.

3. ***Look at your list of coordinates to find the lowest north latitude and the lowest east longitude.*** (*If you are south of the equator, choose the highest south latitude instead; if you are west of 0° longitude, choose the highest west longitude.*) They may be for the same waypoint, or maybe not. Round the numbers for the latitude and longitude coordinates down (*up for south or west values*) to the nearest minute and write these values in the bottom left corner of the frame. These coordinates are the 'point of origin' for the map. Label the bottom left corner of your map with these coordinates. For example, if you had the coordinates as shown in the table on the next page, your point of origin would be 3°55'N and 115°24'E.

4. ***Measure across the bottom of the frame and make vertical ticks for the longitude minutes and seconds*** according to the degrees/distance conversion table in appendix E. For example, in Borneo, on a map of 1:10,000 scale, 1 minute of latitude or longitude is 18.5 cm and 1 second is approximately 3 mm. Use heavier or longer ticks for every 10 minutes and for full degrees.

WAYPOINT	LATITUDE (NORTH)	LONGITUDE (EAST)
W001	3°55'40"	115°24'40"
W002	3°56'24"	115°25'05"
W003	3°55'10"	115°27'15"
W004	3°56'55"	115°26'05"
W005	3°57'45"	115°26'50"
W006	3°58'00"	115°24'50"

5. **Beginning with the longitude at the point of origin, you can count up and label the marks at convenient intervals**—for example, every minute or every 10 seconds.

6. **Measure up the left side of the frame and mark the latitude in minutes and seconds** in the same way.

7. **Find and mark all of your coordinate points on the grid.** Follow the same procedure as for finding coordinate points on a topographic map. If you use a drafting board and a T-square, you can mark the points more accurately and quickly than without.

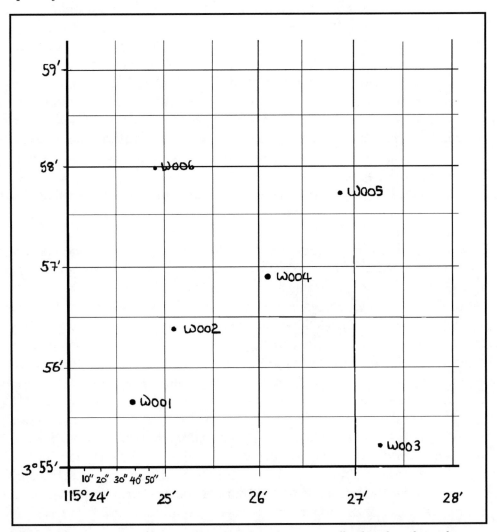

The waypoint coordinates from the table above have been manually plotted on this grid.

11 GATHERING LOCAL KNOWLEDGE ON MAPS

What you will learn in this section:

➤ *A variety of methods for gathering the local knowledge to be drawn on the maps*
➤ *How to make field notes that enrich your information gathering*
➤ *How to interview people using maps; the importance of field-checking*
➤ *Some tips for conducting effective map interviews*
➤ *Whom to interview, and how to include both women and men*
➤ *How to record interviews*
➤ *How to apply the various techniques used in participatory rural appraisal*
➤ *The benefits of photographic cataloguing and how to do it*
➤ *The basic principles of resource inventories*

11.1 TAKING FIELD NOTES WHILE MAPPING

The purpose of surveying in the field is to place local knowledge on the map. You can draw a lot of information on maps during interviews inside someone's home, but it is more accurate to survey in the field. There are two reasons why: the locations are measured, and people's memories are sharper when they see the actual places. Another advantage of surveying is that it gets people out on the land and confirms their connection to the land. Note that although you will collect a lot of information in the field, it is only useful for drawing a map if it is recorded in such a way that you can later read and understand it.

When you are mapping in the field, you are collecting information about land and place. Land information can be described in words (location names or geographical terms), numbers (distances and bearings), pictures, or a combination of all three. It is often easier and faster to describe land by drawing a picture. Numbers add accuracy, and words add clarity. The principle behind taking useful field notes while mapping is to record all three—pictures, numbers, and words—in a systematic way.

There are several advantages to recording notes systematically:

❖ *It allows you to record more information and to do so faster.*
❖ *It makes it easier to draw your map, and to check your notes later if you find a mistake.*
❖ *Systematic notes make it easy for other people to understand your mapping method, giving the map more credibility in the eyes of outsiders. And it makes it easier for other people to work with you on the project or to continue building on your work in the future. (See subsection 9.4.3 and subsection 10.2.3, which describe standardized note-taking for compass surveys and GPS surveys.)*
❖ *Well-organized notes also allow you to easily add to and improve your maps later as more advanced tools and techniques become available to you.*

The key to recording notes systematically is to keep your notebook organized. Here is one way to organize your notebook:

❖ *Prepare a format for taking notes before going into the field.* *You might want to use a notebook or forms prepared on a computer. You may have one format for a compass traverse and another for GPS data (see examples in appendix A).*

❖ *Draw all your sketch maps and drawings on the right-hand pages if you are writing in a notebook.* Write notes about the drawings on the facing left-hand pages.

❖ *Be sure to create enough space in the notebook or on the form to write comments and stories and descriptions of places.*

❖ *At the top of each page write the date, the name of the village or location, and the name of the note-taker.* Later, you and others can tell at a glance when and where you wrote it, and you won't easily confuse it with other papers.

❖ *Write neatly so that others can read it.*

❖ *Write the notes as if they will be a permanent record.* They are originals. Store them safely after the survey, even if you rewrite them or enter them into a computer.

❖ *Write the notes in the language that the local people are most fluent in.* If they are in the local language, the notes can be translated later to the national language if necessary. Using the local language while surveying is important to meet the objective of using the mapping process for cultural maintenance or revival. Also, local people will feel greater ownership of the mapping process.

If you are doing a compass traverse, the points where you collect data are called **stations**. If it is a GPS survey, the points are called **waypoints**. Stations (and waypoints) are numbered. These numbers give you convenient reference numbers for your notes about the places you survey. Even if you are told a story that you can't possibly write in the column on the survey form, you can write the story elsewhere in a notebook with the station number (and name of the place) as reference.

Always record as much information as possible about the place at each station (or waypoint). Try using the *four direction method:* Stand at the station and take compass bearings in four directions: north, south, east and west. Write or sketch what you see in each direction. You can define a given distance around the station, say 20 m, within which you will search for information to record. The distance that you choose will depend on how far apart the stations are.

During the mapping process you will probably be told lots of information that isn't directly related to the maps that you are making—such as the names of plants and their special uses, or perhaps stories and myths about the land. You will have to decide what information is important and how much time you have to record it all. Record as much as possible in notebooks, and also use the other tools and techniques as described in the rest of this section.

Sketches serve as cross-references for the base map and as a quick visual method for taking notes about the landscape. Drawing sketches of your own observations in the field helps you to verify the information on the map. For example, if you are walking and someone tells you about a mountain that you can see, but you don't have the map with you, then draw it in your notebook. Take notes about it, and take bearings to it from a known point. Later you can draw it on the map. Because a notebook is much

SAMPLE GPS FIELD NOTES

WAYPOINT #	DATE	PLACE NAME	GPS COORDINATES	PDOP	AVE. DIFF. (y/n)	NOTES (50 m to N, E, S, W)
001	5/7	Gelengga	113°35'45.8" 1°52'34.9"	2.1	n	N—Rubber plantation E—Mr. Sam's rice field S—Sepik river W—Mr. Sam's rice field
002	5/7	Sasan	113°35'15.5" 1°52'32.7"	1.1	n	N—Sepik River E—Fallow rice field (yr 2) S—Fallow rice field (yr 7) W—Orchard – Mr. Silas
003	5/7	Malanggu (gravesite) Story #26	113°35'26.7" 1°51'45.3"		ave.	N—Sepik River E—Orchard – Mr. Kimo S—Forest fallow (yr. 15) W—Orchard – Mr. Nano
004	5/7	Mt. Kapola Story #27	113°34'2.6" 1°51'10.2"		ave.	N—Sepik R. headwaters E—Carving S—Primary forest W—Primary forest

An example of how to write field notes on pre-prepared forms or in a notebook.

easier to carry and draw on than a large map, it allows you to quickly record what you see, in the form of drawings and notes. If opportunities arise, take bearings where possible, and use the intersection techniques described in section 8.3.4. Later, use the sketch and bearings to determine the location of the feature on the map.

11.2   MAP INTERVIEWS

While walking on the land or participating in field surveying, local people easily remember stories and information about the land. However, if they have never recorded this information before, they may not know what to speak about. The mapping team can help them to talk about the land by conducting a *map interview*. The word 'interview' may sound formal. Actually, we just mean 'asking questions.'

Although doing a map interview while surveying can give you lots of accurate information, you probably don't have the time to walk and survey the whole

Ideally, do map interviews out on the land, where the respondent can point to the place that he or she is talking about.

territory in this way. Moreover, not all community members are able to join the survey teams, so how does their knowledge get onto the map?

Map interviews can also be done inside someone's home. Local people usually know the land so well that they can easily draw a lot of information from memory, straight onto a sketch map or a topographic base map, without surveying. But, having never recorded this knowledge before, they may be unsure about what to draw. Here too, the community mapping team or a facilitator can help to encourage this information to emerge by conducting map interviews, by asking questions.

The following subsection gives you some suggestions for how to ask questions about the land and whom to ask.

11.2.1   WHOM TO INTERVIEW ABOUT THE LAND

Elders are extremely important resource people. They have travelled along the small rivers and trails many times, and they know the area well. In fact, they already have a kind of map in their minds. Your job is to help them to draw their mental maps.

Whom to interview depends on the theme of the map. In every community there are individuals or groups who know a great deal about certain subjects—healers know about medicinal plants, hunters know animal habitats and hunting techniques, elders knows the ancient tales or songs. One or two elders may be able to tell you everything that you need to draw a map of the gravesites for the community. Particular families know about certain geographic areas.

Make a special effort to create opportunities to interview women. There are many reasons that it takes a special effort for women to share their knowledge and views. One reason is that men are often more out-spoken, and women may not talk freely in situations that include men, or not in groups, or not with out-siders. In rural villages, women often work longer hours, and so they may not be available when you want to ask questions. How can you create oppor-tunities for women to participate? Make sure that there is at least one woman on the community mapping team to conduct interviews with women.

Women of some cultures may be more comfortable working on maps in a small group together. Children are pretty well always a part of this activity.

Arrange group discussions for women only, and invite them to bring their children. Adapt the interview schedule to the women's convenience. Speak with individual women while they are doing the cooking or weeding. And, in whatever way, let women know that their knowledge and concerns about the land are unique and valuable.

Don't just interview the most talkative or most educated people. Even in your own community, where you know who is who, it is easiest to interview the most educated people, the ones who can easily understand the map and your questions, and the ones who like to talk. But these people can sometimes be the ones who least represent the community. The quieter people are also affected by what is going on, and they have important knowledge that you might miss by focusing on the loudest.

Group interviews can be effective for learning about the patterns and history of resource use. They are good for collecting a great deal of overview information in a short time and for gathering several perspectives. The dynamics vary, depending on the group and on the subject of the interview. A group situation can assist in keeping the tall tales in check and help to jog memories. On the other hand, some individuals may feel pressure to give a response in front of their peers, even if they don't actually know the answer.

11.2.2 SOME TIPS FOR DOING MAP INTERVIEWS

As much as possible, go to the forest and the fields with villagers and ask them questions there. There is less risk of misinterpreting their answers, and the location provides a stimulus for further discussion. Being at a certain place helps remind the person about the place. Being where something happened can help the person recall memories of the event.

The primary questions in map interviews are Where? How far? How wide? Above? Below? Beside? Right bank? Left bank? North? East? West? South? and so on. All of these questions help you to identify *where*, usually in relation to something else. For example, if an elder is telling you about an important fruit-tree grove, you can help him to find it on the map by asking, *Which rivers is it between? Approximately how far is it from the confluence of the two rivers? How far from the bank of the river? Right or left bank? How far from a rock outcrop (or other such feature)?* Since you are already familiar with the map, use it to direct your questions.

The more complete and accurate the base map is, the more reference points can be used in asking questions, and the more consistent the answers will be. Therefore, the goal of the first discussions should be to put local names and small

The more complete and accurate the base map is, the more reference points there are, and the easier it is to conduct a map interview.

Where possible, visit important sites with the elders who know their location and history.

rivers on the base maps. Probably you will still have to correct names and features as you go along. (Review section 7.2 for a reminder about how to help villagers read a topographic map.)

Find a standard of measurement for distances. Some people who live near logging roads that are marked in kilometres may have a good perception of how far a kilometre is. They will be able to tell you that their swidden is 500 m from a certain river junction. But some people don't really know how far a kilometre is. These people might measure the distance to their swidden in terms of how much time it takes to smoke a cigarette or chew a betel-nut. You might encounter a person who pretends to know how to measure in kilometres but in reality doesn't have any idea. Verify what you are told by walking on the land.

For historical information, estimate years based on significant events. Farmers have remarkable memories about the land that they have worked since childhood. However, they don't always think in terms of years. They may not be able to tell you what year it was when they cleared the forest, or planted a certain tree. But they usually can remember their activities in relation to when the war was, or when a certain king or chief was reigning, or when a certain road was built. The dates of these latter events are recorded elsewhere, so you can estimate the year based on those events.

Cross-check information with several people and using different methods. Trust that local people have complex and in-depth knowledge about the land that they've lived on for generations. But don't expect that they are able to express it in terms that government officials or scientists understand. And don't always assume that they've communicated it clearly, or that you've translated it correctly. For these reasons, it is best to cross-check information. As well as asking people *where*, it's also important to go to the places that they talk about to verify what you have recorded.

Avoid leading questions, ones that can be answered with 'yes' or 'no.' Instead of asking *Do you collect firewood along the Sampanan River?* ask *Where do you collect your firewood?* (You might follow with *Anywhere else?*) Similarly, instead of asking *Is your garden 500 m from the river?* ask *How far is your garden from the river?*

Ask questions in your own words. If you are working with your own people, you know best how to ask questions in your own culture, which words to use. Avoid the use of abstract English words and concepts (for example, 'numerical,' 'trends,' 'evaluation,' 'impact,' etc.). They are often difficult to explain.

Find a comfortable rhythm and pace for asking questions. Rapid-fire question-and-answer may work on TV, but it is not a common way for most people to think or to communicate. Many interviewees would find this style jarring and uncomfortable. In many indigenous communities, people speak in long monologues, answering questions indirectly and at great length. Not everyone responds to questioning in the same way. Some people don't like to talk much and you have to keep prodding. Others like to talk at length, but often go off the topic. You'll find your own way, appropriate for the culture, to keep the interview on track.

Start with broad, open-ended questions. Use the answers to provide clues for formulating more specific questions. Conduct the interview like a conversation. Listen carefully to the answers (recording techniques are discussed in subsection 11.2.3). Ask for backup details and refer backward and forward in time. If you need more information, repeat questions in different ways. If you are not getting the information that you want, change the wording of the questions and try again. From time to time, especially if you are not sure that you understood something, repeat back the highlights of what you heard from the other person and ask for confirmation or correction.

Be flexible and adaptable. Adjust to the schedule of the informants. Develop your own style and accommodate different people and situations. Maybe in the evening in the house you can behave more formally and directly, whereas if you are interviewing someone in the field during the day, you'll have to do more watching than asking.

Finally, show respect, courtesy, and gratitude.

11.2.3 RECORDING THE INTERVIEW

Many of the questions that you ask in a map interview will be worded so that you or community members can draw the answers directly onto the map (as just described, in subsection 11.2.2), but other relevant information will be mentioned at the same time. Notes from an interview can be used to cross-check the information that you draw on the map, and they are also a record in themselves. Use the same notebook that you use to record surveys, or use a separate notebook specifically for each thematic map. Write notes as completely as possible while you are listening to the speaker. Fill in extra details from memory later. *Always include the names of the person doing the interviewing and the person being interviewed, as well as the date.*

If you think that the community members would be comfortable with answering in a more formal interview style, then you could prepare a set of questions on a form. You could create a form for each subject or each thematic map. These forms are easy to file and to refer to. Here too, always record the names of the person doing the interviewing and the person being interviewed, as well as the date.

Interviews can also be tape-recorded if writing is too slow and cumbersome. Be sure to ask permission before recording. Always start the interview by saying the speaker's full name and the date and the time into the tape recorder. On the cassette tape label, write the date, the name of the village, and the names of the key speakers. Two advantages of tape-recording are that it provides the most complete record, and it is less obtrusive and distracting than writing. Disadvantages of tape-recording are that the equipment can be expensive, and it can be prone to technical glitches, such as having the batteries go dead or the microphone get unplugged. In addition, the tapes are time-consuming to tran-scribe, and some words might be hard to

Record the interview by writing notes in a notebook.

understand on the tape, perhaps because of noise. Even if you do use a tape recorder, it is useful to make notes, both as a backup in case of equipment problems and to refer to when asking subsequent questions during the same interview session.

11.3 PARTICIPATORY RURAL APPRAISAL TECHNIQUES

Participatory Rural Appraisal (PRA) is a particular approach to involving the community that incorporates a set of various methods for empowering local people to express, document and analyze land issues and local knowledge. The methods are designed to be simple and fun, like games, to encourage as much participation

as possible. A variety of methods are used so that villagers can cross-check information; knowledge that emerges using one method can be cross-checked with knowledge that emerges using another.

One of the techniques in PRA is sketch-mapping. This handbook already focuses on both sketch-mapping and technical mapping, so in this section we describe other standard PRA techniques (as well as a few additional tech-

PRA techniques are often done in small groups, using available materials.

niques inspired by PRA) that can be used to make the maps more useful. One of the key principles of PRA is that methods are creatively adapted to the local context. Use these examples to get ideas and then adapt or develop your own.

The Game of 'Flags'

Make the activity of finding places on the village map into a game to encourage everyone to participate. This game helps villagers to become familiar with reading a map of their territory. Take a sketch map of the territory or a topographic base map and lay it on a table or on the floor so that a large group can crowd around it. Have plenty of coloured pins or coloured stickers and ask people to use them to mark where their gardens are, or the farthest place that they go to hunt, or where they fish. What they are asked to mark depends on the theme of the map. There will probably be good discussion and laughter as each person takes their turn to place their pins. After everyone has had their chance, and before removing the pins, use coloured pencils to make permanent marks on the map.

The Game of 'Rooftiles'

This game is similar to flags, and its purpose is to show land use during different periods of time. Cut coloured pieces of paper. The size of paper may represent the size of an average farm field according to the scale of the base map, or, if it is a sketch, then the size doesn't matter. Select a colour for each of the years under discussion. As a group, villagers then mark the locations of their gardens using the colour of paper according to the year. If the gardening system is on a rotation, then in some places you will end up with a few pieces of paper of different colours on top of each other. This game can be adapted for different subjects appropriate to the theme of the map.

Activity Calendar

A map shows *where* activities are done on the land, and a calendar shows *when*. A calendar can show when any kind of regular activity is done, such as different farming activities, gathering forest products, fishing, hunting, or sports. The calendar can also show when special annual events occur, such as ceremonies, festivals, or the seasonal closure of fishing. In this way, a calendar complements the maps by completing a picture of the community's way of life.

A calendar can be made for any length of time cycle: yearly, monthly, or daily. A seasonal calendar is usually made for a yearly cycle, but would look different, depending on the culture and the climate of the region. It might show the twelve months of the year, or it might show six seasons named in the local language.

The simplest way to make a calendar is to draw

A calendar shows when *activities occur.*

vertical lines on a large piece of paper, with the time periods written at the top of each column. Then make rows by drawing lines from left to right (or the reverse) and write the activities down the left side, one activity per row. Villagers can use any kind of marker, including coloured pens or coloured paper, to mark what activity they do when.

Matrix

A matrix is a table used to compare the relationship between two or more factors. Or it can help people to understand the relative importance of different products or different land areas to the whole community or to different sectors of the community. A matrix is made in the form of a table, with one factor (and its various components or aspects) listed across the top and another factor (and its various components or aspects) listed down the side. In this way you create an arrangement of square or rectangular boxes, one at each place where the column of a component or aspect of one factor crosses the row of another. In each of the boxes we can put numbers that compare the importance or value or quantity of one factor to another. The villagers are the ones to define the factors or categories, and they also fill in the boxes. Often the values are different for men and women, in which case make two matrices of the same factors, with one to be filled by men, the other by women. The matrices on the next two pages show what types of forest are valuable for which type of forest products.

Matrices are very versatile. Depending on the topic of the matrix, the entries that you put in the boxes could simply be 'x's, or they could be numbers that indicate

Make matrices with small groups so that all can participate. This matrix would enrich a thematic map of forest types by showing which forest types on the map are important for gathering certain forest products.

VALUE OF EACH PRODUCT TYPE FROM EACH FOREST TYPE

	FOOD PLANTS	MEDICINE PLANTS	CONSTRC'N WOOD	WEAVING MATERIAL	CASH	HUNTING
PRIMARY FOREST	●●●● ●●●● ●●● (11)	●●●● ●●●● ●● (10)	●●●● ●●● ●●● (10)	●●● (3)	●●●● ●● (6)	●●●● ●●● ●●● (10)
OLD FALLOW FOREST 15-30 YRS.	●●●● (4)	●●●● ●● (6)	●●●● (4)	●●●● ●●●● (8)	●●●● (4)	●●●● ●●●● ●●●● (12)
YOUNG FALLOW FOREST 5-15 YRS.	●●● (3)	●●●● ●●● (7)		●●●● ● (5)		●●●● ● (5)
FALLOW GARDEN 2-5 YRS.	●●●● ●●● (7)	●●● (3)				●●●● ● (5)
AFTER RICE FIELD 1 YR.	●●●● ●●●● ●●●● (12)					
TEMBAWANG (WILD FRUIT TREE GROVE)	●●●● ●●● (7)	●●●● ● (5)			●●●● (4)	
RUBBER PLANTATION			●●●● (4)		●●●● ●●●● ●●●● (12)	

units of money, kilograms, percentages, hours of work, relative values (or priorities) from 1 to 5 (or 1 to 10), number of occurrences in a given year, area in hectares, units per hectare, number of people involved, number of years of production, etc.

Some other possible uses for matrices include

❖ Showing the relative economic importance of different forest types with a matrix of **FOREST PRODUCTS** and **FOREST TYPE** (filled in with relative abundance from 1 to 5), or a matrix of **AMOUNT OF LABOUR TIME** and **FOREST PRODUCT** (with just one column, to list the number of hours required to produce a certain amount of product)

❖ Showing the distribution of land and wealth in the local economy with a matrix of **LAND-USE TYPE** (rice fields, fallow, cash crops, etc.) and **NAME OF FAMILY OWNER** (filled in with either hectares or units of production or income for each owner)

❖ Showing how land-use patterns have changed over time with a matrix of **LAND-USE TYPE** and **HISTORICAL PERIODS** (filled in with hectares)

❖ Analyzing land-use conflicts with a matrix of **NATURE OF CONFLICT** (trees, land, water, animals) and **DISPUTANTS** or **STAKEHOLDERS** (villagers, neighbouring villages, logging company, government, etc.), (filled in with the number of incidents over the past year, or their relative severity)

RELATIVE VALUE OF ALL PRODUCTS FROM EACH FOREST TYPE

	PERIOD I BEFORE 1947	PERIOD II 1947-1967	PERIOD III 1968-1988	PERIOD IV 1988-1998
PRIMARY FOREST	●●●●●●● ●●●●●●● ●●●	●●●●●●● ●●●●●●● ●●●	●●●●●●●	●
OLD FALLOW FOREST	●●●●●●● ●●●●●●● ●●●●●●●	●●●●●●● ●●●●●●● ●●●●●	●●●●●●● ●●●	●●●●
YOUNG FALLOW FOREST	●●●●●●● ●●●●●	●●●●●●● ●●●●●	●●●●●●● ●●●●	●●●●●
FALLOW GARDEN	●●●●●●● ●	●●●●●●● ●	●●●●	●●●●●●● ●●●●●●● ●●●●
RICE FIELD	●●●●●	●●●●●	●●●●	●●●●●●● ●●●●●●● ●●
TEMBAWANG (WILD FRUIT TREE GROVE)	●●●●●	●●●●●●	●●●●●	●●●●●
COFFEE PLANTATION				●●●●●●● ●●●●●●● ●●●
RUBBER PLANTATION			●●●●●	●●●●●●●

❖ *Showing the consequences of environmental damage to the local economy with a matrix of* **TYPE OF ENVIRONMENTAL DAMAGE** *(erosion, loss of animal habitat, poor water quality, etc.) and* **LAND-USE ACTIVITY** *(logging, commercial fishing, mining, cash-crop farming, etc.), (filled in with percentage damage, estimated values of losses, number of incidents in the last five years, or hectares affected)*

11.4 VISUAL MEDIA

11.4.1 MAKING A PHOTOGRAPHIC RECORD

If you have a camera, take photographs of important places in the territory to use as references in map interviews or to make a catalogue of sites as a permanent document.

Use photographs in discussions with elders who aren't able to walk far on the land. Make notes about the photos during discussions with small groups inside someone's house. For labelling, you might use an overlay of thin paper or plastic, or draw a quick roughly-to-scale sketch of the key features on a piece of paper, or fasten the photo to the centre of a notebook page, and then use arrows from your notes. Use the photo and notes to transfer the information onto a sketch map or topographic map.

Photographs make excellent documentation, if properly identified as to location, direction of view, and date.

Use photos to record and verify the locations of important sites such as gravesites, boundary markings, or culturally modified trees (significant trees marked for ownership or that show evidence of historic use, such as bark or wood removal). Over time, build a catalogue of photographs that show traditional land-use patterns, cultural sites, and important landmarks that demonstrate traditional relationships to the land. Assign a number to each photograph so that you can reference it to a point on the map. Keep the photographs in a book together with the map. Or glue the photos in the margin of the map and draw a neat arrow to the location shown in the photo and indicating the direction of view. This combination helps to validate both the map and the catalogue. In the future, you or someone else can go back to the exact location and take a new photo to show changes over time.

You can use almost any kind of camera. Cameras with a convenient and easy-to-use automatic date imprinter are often available at a reasonable price. A camera that accepts a variety of lenses—such as a single-lens reflex camera (SLR)—is more versatile for taking different kinds of photographs than are simpler models. Learning to use it well can take some time, however, so you may opt for a more convenient rangefinder camera with a wide-angle–telephoto zoom lens instead, or an inexpensive 'point-and-shoot' fixed-focus ('focus-free') unit, or an inexpensive auto-focus model.

No matter what kind of camera you have, here are some tips for taking good documentary photographs:

❖ *Where practical, choose a viewpoint that is easy to identify on the map.* Locate it as accurately as possible by using bearings or error-compensated GPS readings or by reading the contour lines.

❖ *For panoramic sequences, find a spot with minimal obstructions from which to take the photos.* Hold the camera at the same level for each photo (using a support of some kind, or by carefully aligning features seen through the viewfinder). Make sure that there are small overlaps between adjacent photos. For best matching at the edges of adjacent photos (and to best match what the human eye sees), choose a lens that has or can be set to a focal length of around 50 mm (for 35 mm cameras), and mount the photographs so that the transition from one to the next is in the middle of their area of overlap. Uniform lighting (midday with medium overcast) gives the best matching of colours or grey tones from one photo to the next. Note that partial panoramas can also be useful.

❖ *Hold the camera level.*

❖ *Include something that shows scale,* such as a backpack or a person.

❖ *On sunny days, if possible, take the photograph with the sun behind you.*

❖ *It is best if all the important features have about the same amount of light on them.* If some are in bright sunlight and others are in the shade, some parts of the photo may be overexposed or underexposed. Automatic cameras often set the exposure for what is in the centre of the frame, so centre the photo on something of average brightness if practical. If you want to take some pictures in bright sun and others in the shade—especially if you do not have an adjustable or automatic camera—you may want to use a 'fast' (400–1600 ASA) film in the dim places and a 'slow' (100–200 ASA) one in the bright ones.

❖ *Take flash photos only of things that are close by (usually less than 2–10 m, depending on your film, camera, and flash unit),* or they will be too dark to be useful. If the flash goes off automatically when you don't expect it, perhaps you just need to come back when there is more light, but maybe there is something dark in the centre of the frame. If so, recompose the photo and try again.

A panorama photo sequence gives a full perspective of the land.

❖ **Avoid having some important features really close (about 1–3 m) and others far away,** because some of them will probably be out of focus. Keep in mind that automatic cameras usually set the focus for what is in the centre of the frame. It is often best if the focus is set for the closer features.

❖ **Record the following in your notebook:** the date, a few words to describe the subject of the photograph, number of the roll of film (number you give it), number of the photo frame (from your camera), description of the location of the viewpoint, map coordinates, and the bearing to one (or more) key feature(s) in the photo.

❖ **When the photos are developed and printed, immediately number them to match your notebook.** Develop a filing system for the photographs. Label each photo and file it with notes about the location. File the negatives in such a way that you can easily find them to make reprints if necessary.

11.4.2 MAKING A VIDEO RECORD

Video has the advantage of both sound and picture. Use video to make a visual record of certain agricultural practices or ceremonies about the land. As documentation, the video is most effective with some narration. Interview an elder on the topic and record the interview. Depending on your editing expertise, equipment, and budget, you can either include the new footage separately on the same videotape, or you can edit it into the existing footage of the activity. If you have two video cameras, you may choose to use one to record the activity and the other to simultaneously record the interview. But, if you are doing the interview later, you may want to show the video during the interview, so that the elder can more accurately relate his or her comments to what is shown (or has been omitted). File the videos by writing on the sleeve of the video the subjects that it contains and the date that it was recorded.

Videos can have more impact than photographs because of the way that they record movement and sound. And you can watch them right after making them, without having to get them processed first. However, a video camera is more expensive and more fragile than a regular camera and requires more power (batteries or access to the power grid or local source) to operate. Note that videotapes, unlike photographs, need a player and monitor (or TV set) to view them, and they need to be better protected—from dust, heat, moisture, and magnetism. Also, if you want to edit the recordings for a better presentation (by deleting the mistakes and irrelevant parts, or moving sections around), you will need access to additional equipment, or you can pay someone to edit them for you (or perhaps a friendly NGO will be willing to help out).

11.5   COMMUNITY RESOURCE INVENTORY

Among other things, a map can show us where the forest resources are. An *inventory* tells us how much of something there is. When we make an inventory, we measure an area or volume, or count items. A forest inventory is commonly done by company foresters to count the number of commercially valuable trees in an area of forest. Then they can calculate a monetary value for that area of forest or calculate how many trees to cut each year. A rural community, however, values the forest for more than just trees. So, when a community does a resource inventory, the members count not just trees, but all of the plant species that are important to the people in the community. For example, they might include medicinal plants, fruit trees, palm shoots, or vines in the inventory. These aspects of a forest's value are relatively easy to quantify.

Conducting a community resource inventory.

However, the living forest as a whole also has values to a community that are not so easily quantified, such as protecting water and air quality, providing a home for animals, and being a part of the local people's way of life. These less tangible aspects of a forest's value should be described too. So, at the same time as counting, community members should collect information about the characteristics of the forest, such as the soil type, the animal habitats, the slope, and the elevation.

Why count plant resources and collect all of this information? The major reasons are as follows:

❖ *To compare the value of one forest area with another*
❖ *To negotiate compensation for damaged or lost community resources*
❖ *To make a management plan for community resources*

How is an inventory done? Counting every plant in the forest would take too much time. Therefore, we begin by making a list of all the species that we want to inventory. Then we choose small sample areas (plots) and count the number of

Recording the inventory data.

plants of each type on our list that we find in each of those plots. If the location of the sample plots is selected randomly, and we know their size, then the number of plants in the sample plots can be averaged together to give us a good estimate of the number of plants in the whole area.

We locate and measure the sample plots by doing a compass traverse in a grid pattern. In the tropics it is common to do **strip sample plots**. As we walk through the forest, following a straight compass bearing, we count the number of plants of each kind on our list that we see within 5 m to the right and 5 m to the left. When we reach the edge of the area, we come back the other way on the next leg of our survey along a parallel line. In temperate forests, **circular plots** are more common. We measure a given distance and bearing between plots and then mark out a circle of a given radius and count the plants within it. (For more information on how to select sample plots and perform an inventory, you can consult *Participatory Inventory: A Field Manual Written With Special Reference to Indonesia*, by M.C. Stockdale and J.M.S. Corbett, published by Oxford Forestry Institute, 1999.)

As you can see, making an inventory requires many of the mapping skills that are discussed in this handbook. Indeed, one of the products of an inventory is a map of the inventory area, complete with detailed information about the land characteristics. The other important product of an inventory is a table of numbers showing how many of each species are found in the inventory area. If the size and productive potential of the plants is also recorded, then the value of the forest products in the area can also be calculated.

11.6   BUILDING A COMMUNITY LANDS DATABASE

In the process of a mapping project—or even prior to or after the actual mapping—the community may gather a great deal of information about the land that relates to the maps, but that can't be drawn directly onto the maps. This other information could be documented in a variety of forms. The maps and supplementary information could be called a **community lands database**. A **database** is simply a collection of information. A community lands database is focused on the land and could include many different kinds of information collected using the techniques in this chapter, or by third parties as well. You might have a collection of the following items:

❖ *Photographs of land-use types*
❖ *Tape-recordings of elders describing various land-use practices*
❖ *Written stories about the landmarks in the territory*
❖ *A collection of medicinal plants*
❖ *Calendars and matrices about the agricultural cycle*
❖ *Sketch maps of individual farms or hunting areas*
❖ *Resource inventories for particular areas*
❖ *Reports done by academic researchers, or by staff in lands or resources agencies*

Many of these different forms of information can be 'tied' to a thematic map of the community lands, greatly enriching the content of the map. For example, photographs of land uses can be related to a land-use type map. Oral history about landmarks can be related to a cultural map. How can you record, store and organize all this information so that it is easy to make these relationships, say between a photograph and a map?

A simple method to do so is to use a combination of keywords and number codes. A **keyword** is a significant word that is used in indexing a collection of information. For instance, keywords are used by libraries so that people can search for a book by its subject—or subjects, as there are usually several.

Begin by assigning **number codes** to each 'unit' of information. (Think of each story, each photograph, each sketch map, or each plant-collection form as a unit.) It is easiest to store together the units that are the same format. For example, store all of the video-tapes together on a shelf, and file all of the stories in one drawer of a filing cabinet, and file all of the photographs in another drawer. Give each piece of information a number code. The first videotape recorded would be V1, the second one would be V2, and so on. The first photograph taken would be P1 and the second P2 and so on.

Each of these units may contain information about several subjects, some of which can be shown on the map and some not. Suppose that an elder named A. *Lanegan* is recorded on audio-cassette telling a story about an area that his family owns and stewards, and in which there is a *sacred rock* at a place called *Sanit*. He recounts that if a certain kind of cloud is seen at a certain angle from that rock, then the *weather* will turn fair. There is also a rare *Finggela* tree near the rock, he says. Its bark is used as a treatment for *flu*, and its leaves are eaten as a *green vegetable*.

If you were to identify the significant words that describe all of the subjects in this story, you would have a list of words much like the ones that have been italicized above: Lanegan, Sanit, sacred rock, weather, Finggela, flu, green vegetable. In order to turn these terms into useful keywords, you may want to group some of them into categories that extend beyond one example, so your final list may read *Lanegan, sacred rock, weather prediction, tree species, plant medicine, food plant*. Perhaps the day after recording this story, the mapping team asked the elder to take it to his land, and the team recorded the waypoint number, GPS coordinate, and place name. When the written transcript of the story is filed, you would write this information at the top of the file folder:

Number code = **S5**

Waypoint number = **WP4**

Coordinates = **103°25'15"E, 0°15'22"N**

Place name = **Sanit**

Keywords = **Lanegan, sacred rock, weather prediction, tree species, plant medicine, food plant**

The full written story and the details of the mapping data would be filed in the folder. Be sure to record in the folder the number for the audio-cassette from which the story was transcribed.

Someone who looks at the map and becomes curious about a particular place can search in the files for a picture or story about it. For example, if this person wants to know more about the site marked by 'WP4' and a symbol of a *sacred rock*, a trip to the files to find 'WP4' will reveal the rest of the story. You can see that it is possible to find the place on the map from the story in your files—or to find the story in the files from the map.

The story also contains information about *weather prediction*, which is not directly related to the map. With this method of organizing the data you can still search your database for information that isn't drawn on the map.

You can label photographs and other information units with number codes, way-point numbers, etc. as well.

Notice that one way to relate your information units to the maps is to make sure that all of the words that you use in the legends for the thematic maps are made into keywords. In this case, *sacred rock* is in the legend of the map of sacred places and *Lanegan* is in the legend of the map of land ownership. Therefore, the place discussed in this story can be found on the map of sacred places by the waypoint number, the place name, and the keyword *sacred rock*. It is found on the map of land ownership by the waypoint number, the place name, and the keyword *Lanegan*.

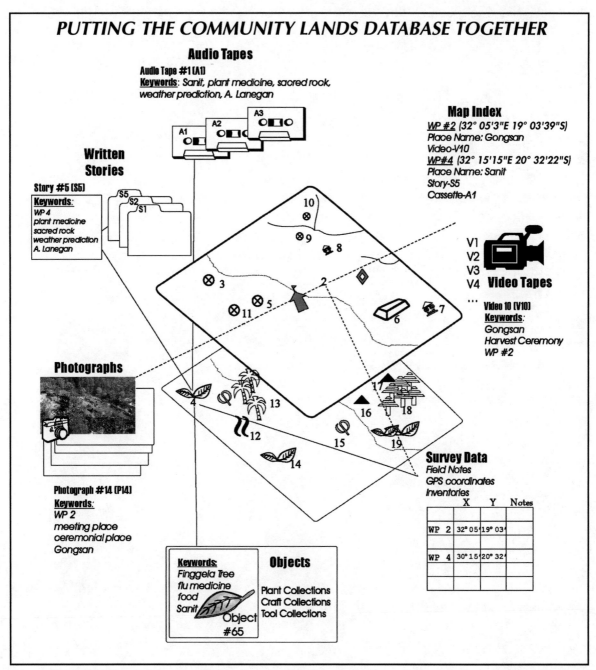

PUTTING THE COMMUNITY LANDS DATABASE TOGETHER

Audio Tapes

Audio Tape #1 (A1)
Keywords: Sanit, plant medicine, sacred rock, weather prediction, A. Lanegan

Map Index
WP #2 (32° 05'3"E 19° 03'39"S)
Place Name: Gongsan
Video-V10
WP#4 (32° 15'15"E 20° 32'22"S)
Place Name: Sanit
Story-S5
Cassette-A1

Written Stories

Story #5 (S5)
Keywords:
WP 4
plant medicine
sacred rock
weather prediction
A. Lanegan

Video Tapes

Video 10 (V10)
Keywords:
Gongsan
Harvest Ceremony
WP #2

Photographs

Photograph #14 (P14)
Keywords:
WP 2
meeting place
ceremonial place
Gongsan

Objects

Keywords:
Finggela Tree
flu medicine
food
Sanit
Object #65

Plant Collections
Craft Collections
Tool Collections

Survey Data
Field Notes
GPS coordinates
Inventories

	X	Y	Notes
WP 2	32° 05'	19° 03'	
WP 4	30° 15'	20° 32'	

The various components of the completed community lands database fit together to give a comprehensive picture of the community and its lands.

If you want to make access by keyword even easier, so that you don't have to flip through all of the folders to find the one that you are looking for, you can compile an *index*. Once you have completed all of your map interviews, make an alphabetical list of all the keywords that are used on any of the files. After each keyword, list all of the numbers for the information units (stories, audio-cassettes, videotapes, photos, etc.) that refer to that topic. Place names can either be included in the same index or compiled as a separate list. Make a similar list for waypoint numbers.

If you have the time, you can add cross-references to make your database more usable. For example, you might enter into your index 'clouds. *See* weather prediction,' 'elders. *See* Lanegan' (include a complete list of elders referred to in the database), 'Finggela. *See* tree species,' 'flu. *See* plant medicine,' and 'vegetable. *See* food plant.'

⑫ PRODUCING THE FINAL MAP

What you will learn in this section:
➤ *How to organize and compile different types of survey data*
➤ *How to design the final map*
➤ *What the essential elements in a map legend are*
➤ *How to choose a method to reproduce maps*
➤ *What drafting materials and equipment are necessary*
➤ *How to apply lettering, patterns, and lines*
➤ *How to draft final maps, step-by-step, from the field maps*
➤ *The importance of checking back with the community*

12.1 COMPILING THE FIELD SURVEY DATA

This chapter is about how to produce final maps from the survey data that the mapping team has collected with the community. At this point the community has produced

❖ *Sketch maps drawn by community members*
❖ *Topographic base maps (if available) with local knowledge roughly drawn on them*
❖ *A compass survey or GPS survey plotted on graph paper, along with survey numbers, sketches, and notes*

How do we put all this information together to make a professional-looking map that is easily understood by villagers and outsiders alike? First, you need to compile and clarify that information.

12.1.1 COMPILE AND CLARIFY THE FIELD SURVEY DATA

Compile (collect) the community data in one place. For mapping projects that involve more than a few days of field-mapping activities and those that include several people's drawings and notes, it is especially important to gather all the field notes together in one place, along with the rough maps.

Check all of the notes for completeness and consistency. Cross-check inconsistencies in the notes within the survey teams and between survey teams if you haven't already. Although it is best for all the team leaders or note-takers to check through the notes together at the end of each day while it is still fresh in their minds, there are many reasons why that doesn't always happen. For instance, perhaps one survey team campedin the forest during their survey and then there was a wedding ceremony in the village, so mapping work ceased for the next two days. Therefore you need to compile and check all of the notes and rough maps now, before you can make the final map.

You can rewrite the notes in clean notebooks if you choose, or enter notes into a computer. But it is best to keep reasonably neat and organized notes in the first place so that you don't need to do this extra work before you can proceed. You want to list the GPS coordinates in the legend of your reference base map (final base map), so you might as well type them into the computer if you have one, or else use a typewriter, if available, so that they are legible and neat.

12.1.2 PROCESS THE FIELD DATA

The field data must be 'processed' (plotted) before using it to draw the final maps. Compass data must be plotted on graph paper (review section 9.5). GPS coordinates must be plotted on a topographic map or on graph paper (review section 10.4). All the local knowledge collected in the field with the survey data should be drawn on rough maps using the appropriate field symbols.

Local information collected through interviewing—whether on sketch maps, topographic base maps, or 3-D models—should be compared with the results of the survey data. Check that the locations are correct and the names are correct and consistent. Because they have been measured, the surveyed locations should be trustworthy, but the sketched and estimated information is also valuable for cross-checking place names and stories.

Ideally, much of this data processing was done with groups of villagers each evening after surveying in the field, because the best time to do so is when the information is fresh in people's memories, and the additional information and cross-checking improves the focus of the next day's survey and data collection. But perhaps (as often happens) large groups of people attended the evening sessions and a great deal of information was revealed at once, sometimes to the point of confusion. Notes had to be taken quickly and roughly. Now you need to refine the data processing before drafting the final maps. For example, locate the GPS coordinates again, but this time more carefully and precisely than was done during the earlier session.

12.1.3 SELECT WHAT TO TRACE ON EACH THEMATIC MAP

The schematic on the next page shows a typical scenario where survey data and sketch maps of local knowledge are processed with a common topographic base map before the information is split onto separate thematic maps. Notice that the rivers, mountain peaks and major roads are the common information traced from the base map onto all of the thematic maps.

Part of organizing the data and designing the thematic maps is to decide what information will be drawn on which thematic map. Once the data is processed and the rough maps are compiled, you can combine or split information onto separate thematic maps in any way you want. Use tracing paper to select and trace information from any of the field maps. If the information on the field maps is cluttered, then you may want to trace different categories of information onto different maps. For example, put the *hunting* information on one map sheet, the *sacred sites* on another, and *land tenure* on a third. Alternatively, you may want to summarize information by taking information from two or more maps and putting it onto one. Of course, every time you copy information from one map to another, be sure that each feature is precisely positioned, and methodically check to make sure that all the necessary features have been copied over to the new map. (Review the sidebar 'Generalization' at the end of section 2.2, and generally follow the drafting techniques described in subsection 12.3.2—but without the inner frames, and don't use any ink or stick-on materials yet!)

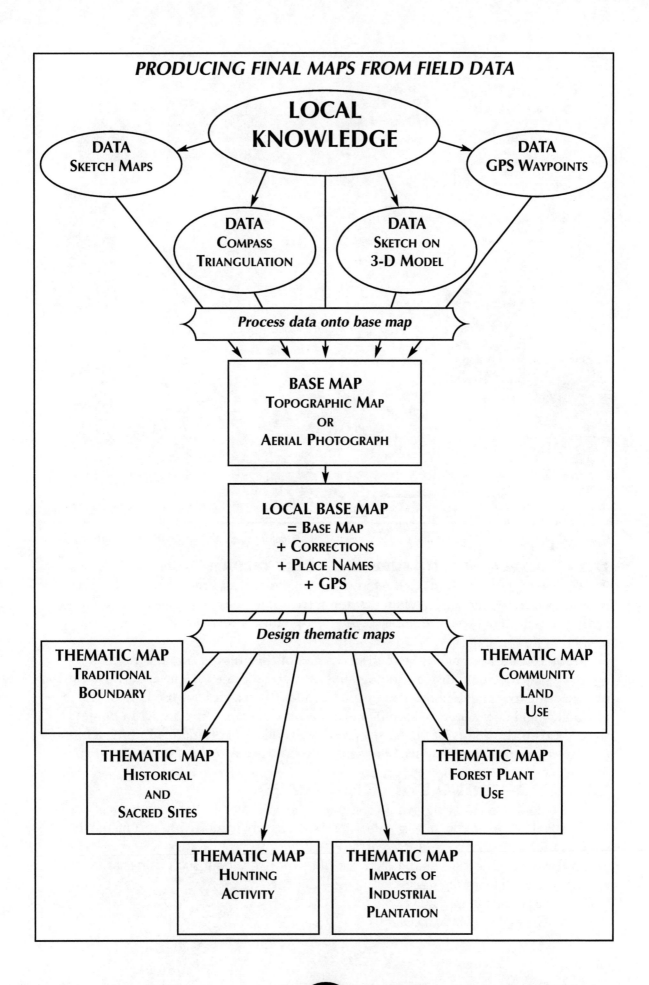

PRODUCING FINAL MAPS FROM FIELD DATA

LOCAL KNOWLEDGE

DATA SKETCH MAPS

DATA COMPASS TRIANGULATION

DATA SKETCH ON 3-D MODEL

DATA GPS WAYPOINTS

Process data onto base map

BASE MAP TOPOGRAPHIC MAP OR AERIAL PHOTOGRAPH

LOCAL BASE MAP = BASE MAP + CORRECTIONS + PLACE NAMES + GPS

Design thematic maps

THEMATIC MAP TRADITIONAL BOUNDARY

THEMATIC MAP COMMUNITY LAND USE

THEMATIC MAP HISTORICAL AND SACRED SITES

THEMATIC MAP FOREST PLANT USE

THEMATIC MAP HUNTING ACTIVITY

THEMATIC MAP IMPACTS OF INDUSTRIAL PLANTATION

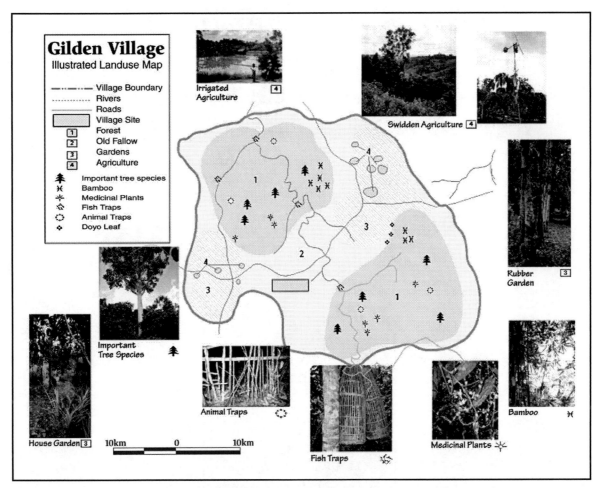

Displaying photos around a map is one way to make a map both more interesting and more informative.

12.1.4 DISPLAYING THE SUPPLEMENTARY INFORMATION

A simple system for filing and referencing supplementary information to the maps is explained in section 11.6. When designing the final maps, consider what number coding system to use to correspond to your system for filing and cataloguing the supplementary information.

In some cases you might like to display some of this information on the final map. For example, you could display photographs of each land-use type around the perimeter of a land-use map. Each photograph would be identified by a land-use type in the legend and would correspond to specific polygons on the map. Or you might choose to write short stories about particular sites in the legend of a cultural map, using a number code to relate each story to the appropriate site on the map.

12.2 DESIGNING THE FINAL MAPS

The mapping team or the whole community made some initial design decisions before starting the field mapping—decisions such as what information to map, what scale to use, and what symbols to draw on the field maps.

At this stage there are final design decisions to make about such things as

❖ *The size of the map sheet*
❖ *The position of the legend*
❖ *What information to include*

❖ *The symbols to use*

❖ *The size and style of lettering, etc.*

Recall that all maps are merely simplified representations of the real world. It is the map designer's job to select and generalize the information so that the map can be used effectively to communicate something about the land. Cluttering the map with unnecessary information, or emphasizing the wrong thing, can give a false impression of its meaning and confuse the person reading the map.

It's satisfying and beneficial to produce a good-looking map, but it's even more important that it communicates information well. To achieve this goal, the cartographer(s) must be clear about what the map is to be used for, who will be reading it, and why.

A good map should be

❖ **Easy to understand**

❖ **Meaningful** *and* **relevant** *to the map reader*

❖ **Suited to the needs** *of the map user*

❖ *Sufficiently* **detailed** *and* **accurate** *according to the purpose of the map to give a true picture*

❖ **Clear**, **legible**, *and* **interesting** *or attractive to look at*

How do we achieve these criteria? One way is by considering the visual impact of the various elements on the map. By changing the size, the colour, the tone, and the shapes of symbols and lettering, as well as the contrast between symbols and lettering, the cartographer can give more or less emphasis to selected information (see subsection 12.2.2). At another level, the size and the placement of the title, legend, and scale affect the reader's focus of attention and his or her impression of the map (see subsection 12.2.4).

12.2.1   DECIDE ON A METHOD TO REPRODUCE THE MAP

Decisions about symbol design, map layout, drafting materials, and technique all depend on the method that you choose to reproduce the maps. There are two basic methods for reproducing hand-drawn maps: photocopying and blueprinting. Commercial printing—black-and-white or colour—could be considered if you need many copies, such as for posters. Maps can also be drawn on computer using special software and printed using a piece of equipment called a *plotter* (or, for small maps or pieces that can be tiled together, even on a regular printer).

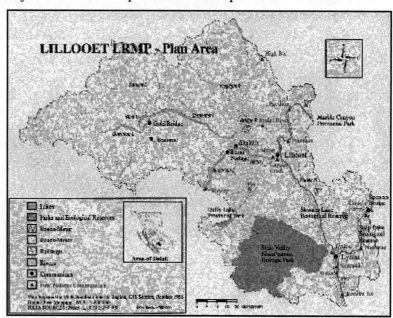

An example of a map from British Columbia, Canada.

Blueprinting is the cheapest way of reproducing a large number of map sheets. It is a widespread technology that has been around for a long time. The blueprinting process works with translucent paper only, so your original must be drafted on tracing paper or plastic film. Drafting a map for blueprinting takes careful planning, because it is difficult to correct any mistakes. You can't use 'Wite-out' (or other opaque correction fluids), and erasing often leaves a mark that shows up in the copies.

With blueprinting, the design options and choices of materials are more limited than with photocopying. For example, you can't use a computer to write and print a legend to glue onto the map sheet (unless you print on special transparent adhesive paper or plastic). Though you can make blueprints of pencil-drawn maps, ink makes a sharper copy. A blueprinting machine needs careful adjustment, and the original needs to be clean if you want to get a clear reproduction.

Photocopying large-format maps is more expensive than blueprinting, but, as the technology improves and expands, it is getting cheaper to photocopy. Photocopying usually makes cleaner-looking copies than blueprinting—depending, of course, on the quality of the machine. Most cities have at least one place that does large-format photocopying on quite-new machines. Photocopying allows for more flexible drafting techniques, because the process does not depend on transparent paper (though tracing paper can be used). Mistakes can be corrected with correction fluid. You can use a computer to print the symbols and legend on regular paper and glue it onto the map.

Colour photocopying is available in specialty printing shops, but you may find it to be very expensive. As you can see, it is best to design your maps for black-and-white reproduction. You may want to colour your maps later by hand, or pay for one or two expensive colour copies, but you always want the option of making several less expensive black-and-white reproductions. Therefore, it is best to use symbols that are evident in black-and-white, to avoid the confusing product that results from putting a coloured map through a black-and-white copier, in the process losing the meaning of the map as contrasting colours turn into near-identical greys. The best compromise is to try to choose black-and-white symbols (shading patterns, lines, and point symbols) that are conducive to colouring by hand.

Another way to produce and reproduce maps is by using a *computer*. A map is drawn (*digitized*) into the computer using a device known as a *digitizing tablet* and stored electronically. Maps are printed from the computer in colour on a machine called a *plotter*. Multiple copies can be made. Once the map is in the computer, it is easy to change any aspect of the map. You can add and update information. You can easily experiment with the design by trying different symbols, lettering and colours. And there is a huge variety of symbols and lettering to choose from that are all easily reproduced as long as you have the equipment. Computers create special design challenges and opportunities that are beyond the scope of this handbook, however. Therefore, the following sections will focus on designing maps for hand-drafting.

12.2.2   REFINE THE SYMBOL DESIGN

While doing the field surveying, you jotted notes down quickly, perhaps with abbreviations, and you recorded features on the field maps with quick, simple symbols (perhaps with colour pencils). Now you need to refine the symbol design for the final map in accordance with the chosen method for reproducing the map and the drafting materials available.

Design the legend first, using a clean sheet of paper. It will be a reference sheet for the symbols. Include additional reference information, such as font styles and sizes for lettering, pen sizes for lines, and template numbers for symbols. Making a reference sheet is useful for testing out the design, and it minimizes mistakes as you draw the map. It is also useful to maintain consistency if there is more than one person working on the map, or if someone needs to make changes in the future.

Note: For best results, maps that you intend to reproduce by black-and-white photocopying or blueprinting should be drafted strictly in black. Greys generally do not copy well, often either turning black or disappearing. If you do want something to appear grey, you have to draft it onto your master map as a close arrangement of tiny black dots (it looks grey to the eye). There are at least three strategies: you can cut the shape out of a commercial stick-on (or rub-on) sheet (ask about the different **dot screens** at a graphic supply store), you can use a computer and software to print out **halftone** images, or you can draft all the grey items on a separate piece of paper or mylar (see subsection 12.3.1) and talk to a commercial printer about creating a halftone copy.

It is useful to think of three different categories of features: **point**, **line** and **area**. Each category is shown with a different grouping of symbols. Consider how the symbol design creates *contrast* between the features, adding clarity to the map. Consider which features you want to make stand out more than others by using contrast.

Point Features

Point features are things that are too small to be drawn to scale on a map and still show. Depending on the scale of your map, they may include gravesites, houses, sacred trees, wells, honey trees, fruit-tree groves, bat caves, bridges, gardens, animal dens, etc. They can be shown with small pictures, abstract designs—such as squares or circles—numbers, or letters. Small picture symbols are the most meaningful to community folk and can be used to produce an aesthetically pleasing and interesting map.

As an example of the use of point symbols, let's consider that you are using a 1:20,000 scale map, and you want to represent a house 10 m wide with a small picture symbol. You want it to be big enough to see easily, so you draw it 5 mm across. In reality, the space covered by that symbol represents 100 m on the ground, which is 10 times the actual size of the house! However, as long as everyone understands that the symbol has been exaggerated in size, there won't be any confusion.

In fact, it is standard practice to use a uniform size of symbol for all of the point features of each kind—such as houses, or sacred rocks, or gravesites, etc.—regardless of the feature's actual size. However, sometimes the size of the symbol can give additional information. For example, on a small-scale map, a large square could indicate a large town and a small square could indicate a tiny village.

One thing to be aware of is that the use of bigger-than-scale symbols may interfere with other features that you want to show on the map. For instance, you may have several houses side-by-side just a few metres apart. What can you do if each 10 m house appears to take up 100 m of space on the map? Well, if most of the map is not that crowded, you can just move each house symbol a little bit away from its

Examples of abstract point symbols and picture symbols.

correct position, so that each one can be clearly seen. (Review the sidebar 'Generalization' at the end of section 2.2.) An alternative would be to invent a new symbol to represent a group of houses. Or you might decide that the whole map is too cluttered and should be redrawn at a larger scale. Or maybe you just need to draw a separate larger-scale map for the crowded area, with a note on the main map to see the other map for more detail.

Keep in mind the scale and purpose of the map when deciding how big to make point-feature symbols. If extreme accuracy is the highest priority, you will want to make the symbols smaller than if it is more important that the map be easily seen by a group of people at the same time. In general, draw symbols as small as possible while making sure that they are still clearly differentiated and easy to see.

Point symbols can be drawn freehand or traced using a template. If drawing the pictures freehand, the challenge is to make sure that they are all drawn identically so that the map reader knows immediately that it is the same symbol, and so that the map is neat. Obviously, the simpler the picture, the easier it is to redraw identically. Another technique is to use adhesive symbols. You may want to design symbols with a computer; print on adhesive paper with a removable backing or use regular paper and glue.

Line Features

Line features are things that are long and thin, such as roads, trails, rivers, and boundaries. Use different line patterns for representing each of the line features on the map. Lines can be drawn solid or as dashes, 'x's, or squiggles. The thickness of a line usually indicates its importance on the map. You can heighten the contrast and make certain lines stand out by drawing them thicker or heavier. This tactic helps the reader to differentiate the lines and conveys meaning by visually showing their relative importance.

As with point features, you may find that you need to exaggerate the thickness of line features to make the symbols stand out clearly.

You can help make your map readable and save yourself some design work if you remember that many line features have standardized symbols. Look at existing maps that you have access to. Notice that paved roads, for instance, are usually depicted with solid heavy lines or a pair of parallel lines; black or red is often used. Rough roads are shown with a solid dashed line or parallel dashed lines. Trails are marked by thinner dashed lines. Rivers are indicated by solid lines of grey (blue if it is a colour map) or by parallel lines with the area between coloured in with a lighter tint. Boundary lines are usually some combination of dots and dashes, though 'x's or squiggles can be appropriate too.

You can draw lines freehand with a good technical pen. A trick for making crisp dashes (this technique works best on mylar—see subsection 12.3.1) is to gently scrape the dried ink off squarely, using a very sharp knife. You can also make lines by buying and applying adhesive tapes that are printed with various patterns. These tapes are much better for straight lines than for tight curves; crepe-type tape will bend better than other kinds.

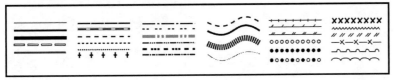

Sample line symbols that you can draw by hand or with a computer.

Area Features and Polygons

Area features are things that are large enough to be shown to scale on the map. Depending on the scale of your map and the actual sizes of the features, area features can include such things as rice fields, mangrove forest, meadows, lakes, hunting areas, deforested areas, and so on. (If the map is at a very large scale, say over 1:5000, you may wish to consider representing such things as roads and buildings as area features drawn to scale.) Area features are shown as polygons on a map.

A *polygon* is simply an enclosed area with a defined boundary. Polygons can be shown with colour, various black-and-white patterns, or with an outline (type line) with a code inside it. Polygons on a map are almost always next to other polygons and may even overlap them, so the key in designing area symbols is to use contrasting textures or colours to help the reader's eyes quickly differentiate one from another.

Colour is the easiest-to-read way to distinguish area symbols on a map. It is easier for the human eye to differentiate colour than either shades of grey or black-and-white patterns. Consider in your choice of colours that humans can react to colour in an emotional way or from cultural connotation. For example, in many cultures red connotes warning or danger or damage or heat. Green connotes vegetation and health. Yellow or brown often symbolizes dryness or soil. Blue is a universal symbol for water.

Another thing to keep in mind is how colours appear in relation to other colours. Some colours, when placed beside each other, contrast well and are easy to differentiate, whereas some aren't. Also be certain that the colours you choose will not be mistaken for each other when they are widely separated, with different neighbouring colours (unless the the context will make it clear which is which). For example, a given shade of light green will look lighter when surrounded by dark brown than if it is beside light yellow. Remember too that some colours, contrasting or not, are not aesthetically pleasing when we look at them together.

Also note that about 8% of men and 0.5% of women on Earth are to some degree colour-blind, especially when it comes to distinguishing red from green. Therefore it is advisable to confirm that everyone in the community who needs to will be able to distinguish differently coloured symbols and not confuse them with each other.

The advantage of black-and-white patterns as polygon symbols is that the reproduction is cheaper than for colour. Black-and-white patterns could include parallel or crossing lines, patterns of dots, or patterns of representational symbols. For clarity and neatness, a pattern must be consistent. If you fill an area with dots or lines, for example, they must all be the same size, in the same pattern, and at the same spacing. Working freehand, the simplest patterns consist of ruler-drawn lines. Still, it can be very time-consuming and difficult to make the pattern consistent and neat.

A more professional-looking but more expensive way to fill polygons is to buy sheets of preprinted patterns. The patterns are printed on sheets of adhesive-backed clear plastic film. To apply them, simply cut a portion of the sheet to the shape of the area on your map, peel off the backing material, and then place it on the map. A similar selection of patterns is available as rub-on transfers but, like rub-on lettering, they are less durable than the adhesive type.

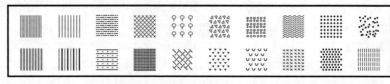

Sample area patterns that you can draw by hand or with a computer, or that you might apply from a self-adhesive or rub-on plastic sheet.

Important: Place stick-on patterns on the map only after you've finished drawing with pen, because ink won't stick well to the plastic, and it will seep underneath every time you draw across or against an edge.

If you have to place the pattern over lettering on the map, then cut a small hole in the pattern or else the lettering will be difficult to read.

Patterns should be simple and repetitive so that they are easy to see, even over a small area. Consider how two different patterns will look if drawn adjacent to each other. Are they easily differentiated? Patterns must also be distinct from any line or point symbols on the map.

Polygons can also be depicted with **type lines** (out-lines), with a number or letter code in the centre of each polygon. This polygon symbol is visually the most difficult to read on a map, because it is easy for the reader to confuse the type lines with rivers or roads, and because areas of colour or pattern are more quickly spotted and recognized than are codes in small type. However, this type of symbol is good to use if there are many different polygon types on one map, because there an infinite number of different codes to choose from, and we don't need to depend on our eyes to differentiate the colour or pattern from others that we might perceive to be similar. We simply read the legend to interpret the codes.

Type lines are also good to use for showing overlapping areas, for example to show overlapping boundaries, because it can be difficult to overlap pattern symbols or colours.

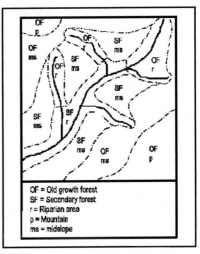

One way of identifying polygons is with a type line around each polygon and a number or letter code inside. A standard format for type lines is a dash-and-dot sequence.

12.2.3 LETTERING

The **lettering**, also known as **annotation** or **typography**, can be done in various ways. Hand-lettering takes skill and practice to do consistently and neatly. Alternatively, you can trace letter templates with a technical pen. Sturdy templates will last a long time, so they are an inexpensive way to produce consistent lettering. Preprinted rub-on lettering is also widely available, easy to use, and looks good, but it's expensive, and the letters can crack, chip or scrape off your map after application.

You can also use a computer to print lettering. If you are blueprinting the map, then you need to print the lettering on special adhesive-backed clear plastic film. But if you are photocopying it, print the lettering on regular paper. Then cut out the labels and place and glue them onto the map.

Choose a lettering style that is clear and legible in both small and large print. A map usually looks neater if you use the same letter style (font style) throughout. But it is also helpful to vary the style (by using *italic*, **bold** or ALL CAPITALS, for example, or serif and non-serif fonts). You want to make it easy for the map user to distinguish waterbody names from the names of sacred rocks and the names of garden owners, for instance. Choose a style that is compatible with the content of the map. Some styles are more formal or less formal than others. For clarity, it is usually better to choose conservative styles. Use bold text only to emphasize those elements that you want to draw attention to.

On the reference sheet, specify the letter style and the size for labelling every different kind of feature. Be clear about which features are to be labelled with large letters and which with small letters. Whether to use large or small letters depends on the relative importance of the different types of features. For example, on a thematic map that shows land ownership, the family names identifying the polygons could be bigger than the names of the rivers. For legibility, all text should preferably be at least 10 or 12 points (capital letters 2–3 mm in height) and certainly not less than 6 points.

To enhance the readability and neatness of your map, follow the cartographic conventions for the placement of labels. Where possible, place labels so that the base of each letter is parallel to the bottom of the map sheet, making the labels easy to read when looking at the map with north at the top. Point features should always be labelled in this way. Labels for area features are normally written this way too, but if an area is too small for the label to fit, it can be curved to fit the polygon. If it still won't fit, you may need to put the label outside the polygon, with an arrow to indicate to which polygon it refers.

The names of linear features (such as roads and rivers), however, are customarily written along the feature instead. If the feature runs directly north–south, then place the lettering to be read from top to bottom, or from the left margin. If the line curves, you can make the lettering curve too, but it should not be so curved that the letters are wildly twisted. Strive for a smooth flow, or just slant the whole label (or each word in it) so that the text is at the average angle of that part of the line.

If you find that the map is cluttered with letters all crunched together, consider redrawing the map at a larger scale. Another option is to use number/letter codes on the map itself, with the bulk of the information in the legend.

Never place any lettering upside down either in whole or part. No letter should ever be any more than 90° from right-side up.

CAPITAL CITY
CITY NAME
Town Name
Village Name
Special Feature
Miscellaneous Comment
Small Lake
BIG LAKE
River Name
NAME OF REGION
(Fancy Text)

Choose easy-to-read, attractive fonts for the lettering. Vary the typestyles slightly (size, bold, italic, capitals) to label different kinds of features.

6 point Helvetica	6 point Times
8 point Helvetica	8 point Times
10 point Helvetica	10 point Times
12 point Helvetica	12 point Times
14 point Helvetica	14 point Times
16 point Helvetica	16 point Times
18 point Helvetica	18 point Times
20 point Helvetica	20 point Times

Examples of two common typefaces set at eight different point sizes.

12.2.4 DESIGN EACH OF THE MAP ELEMENTS

Other map elements, besides the thematic symbols, are essential to inform the reader about how to read the map. Like the symbols, these elements should also be designed first on a separate sheet of paper. A reference sheet is especially important for these elements, because many will be duplicated on each map in a series. To make a map series consistent, make these elements identical in design and placement.

Map title: Choose words that tell the reader immediately what the map is about. For instance, include the thematic subject and the name of the community. Decide which language to write the title in. The reference sheet is the place to design and record the size and style of the lettering for the title.

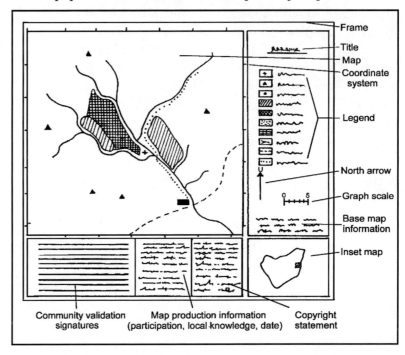

Community validation signatures — Map production information (participation, local knowledge, date) — Copyright statement

Frame — Title — Map — Coordinate system — Legend — North arrow — Graph scale — Base map information — Inset map

Map series number (optional): If the community made a series of maps of the same area, then give each map a number. This series number helps for filing and reference.

North arrow: If the map is for display and discussion purposes and will not be needed for navigating on the ground, then you only need to indicate true north. You can use a ruler and other tools if you want to design a fancy arrow. If the map might be used for navigation or further mapping, then indicate declination also—an easy way is by photocopying the north arrow from the topographic base map.

Graph scale: Always draw a graph scale on every to-scale map, so that even if the map is reduced or enlarged on a photocopier, anyone who needs to can still always figure out the scale.

List of symbols: Arrange the order of the symbols listed in the legend. List the point symbols together, the line symbols together, and the polygon (area) symbols together.

Location map: This small (and small-scale) map of the region shows where the area of your community map is situated in relation to other features in the region (often in relation to the rest of the country, state, or province). Show only the major rivers, highways, cities, coastlines, boundaries, etc. Draw a frame on a small piece of tracing paper

---

ESSENTIAL ELEMENTS ON EVERY SCALE MAP

☞ Title

☞ North arrow

☞ Graph scale

☞ Legend of symbols

☞ Location map

☞ Date when the map was produced

☞ Base map reference

☞ Survey information

---

and within it trace the necessary information from any small-scale map that you have of the region. Keep it simple. Then use shading or a heavy outline to show the location of the community map within this area.

Coordinate system (optional): If there is specific information about the coordinate system written in the legend of your base map, then write it also on your map. Don't obscure your map by drawing many lines across it—just draw a tick and note the grid line value in the margin of the map. There should be at least one such mark, and preferably two, for longitude and for latitude (or easting and northing).

Base map reference: Identify the base map you used. If it was a topographic map, write down the series name and number, map sheet number, date of publication, and the name of the organization that published the map. If it was air photos, write down the photo series number and date.

Survey information: If the map was made from a compass or GPS survey, record that fact, as well as the date surveyed and the names of the surveyors.

Source and method of the thematic information (optional): Write one sentence about where the thematic information came from. For example, 'The thematic information on this map is from the local knowledge of the people of Nade village.'

Copyright (optional): Write a copyright statement that the information on the map is owned by the people of the community. What does 'copyright' mean? Once you have designed your own symbols and located features in the field and drawn them onto a map, then you have the copyright for your map. If something is 'copyright,' it means that other people are not allowed to reproduce the map without your permission. Copyright applies not just to the design of the map, but also refers to the unique traditional knowledge that is used to draw the map. It means that communities own the maps that they make and they have the right not to allow the maps to be photocopied without their permission.

Signatures of validation (optional): The community may want to create a space on the map for signatures of validation. People who, in the eyes of the community members, have the authority can sign here to show that they approve of the map. These people could be elders, the village headman, the mayor, the village council, or the tribal council.

Explanatory note (optional): Sometimes it is useful to write an explanatory statement about the purpose of the community map, who made it and what for.

EXAMPLE OF A COPYRIGHT STATEMENT

This map was made and validated by the people of _____ village. Residents of _____ village hold the copyright for this map and the information on it. Any use or reproduction of this map must be with the consent of _____ village through a decision of the village council.

EXAMPLE OF AN EXPLANATORY NOTE ON A COMMUNITY BASE MAP

This map is the base for a set of three maps depicting the traditional boundary, history, and land use of the people of Long Lamai. It is part of a larger map series made by and for Penan communities of the upper Baram and Limbang Rivers. We Penan have made these maps to record and communicate our culture and our use and occupancy of these rivers and forested lands since time immemorial.

Disclaimer: Most community maps should contain disclaimers that explain that the map is incomplete, pending further data collection and research. That way you leave an opening for making changes, and you have a way out if someone tries to claim to the disadvantage of the community that what the map says is completely accurate, when it has become apparent that it isn't quite accurate, or that things have changed since the surveying was done. Indeed, data collection should be an ongoing project.

What language will you use for these map elements? It depends on the purpose of the map. If the maps are to be used as a documentation and educational tool for local people, then it is important to use the local language. If the maps are also to be used to communicate with government agencies and outsiders, then you'll need to use both the local language and the national language. Sometimes you'll also be communicating internationally, for example, to funding agencies or at international meetings. Then English or another appropriate foreign language could be added.

12.2.5 ARRANGE THE MAP LAYOUT

The final step in designing the maps is to arrange the placement of all the components of the map within neat frame lines. Lay the rough map, which you will be tracing, flat on the drafting table. Place the tracing paper on top and adjust the paper to leave room for space between the map border and the map and for placement of the other map elements—the title, the legend, the location map, etc. Arrange the placement of the map elements, aiming for a neat and harmonious appearance. Again, consider which elements should have more or less prominence according to their importance in regards to what the map is intended to communicate.

Elements on the map should be balanced and relatively symmetrical. Imagine a visual centre point slightly above the real centre of the page. It is a pivot around which the map elements are balanced. The map elements are each visual objects that have two characteristics: *weight* and *direction.*

Visual *weight* works like this:
- ❖ *Whatever first attracts your eye is heavier.*
- ❖ *Large objects appear heavier.*
- ❖ *Red looks heavier than blue, black looks heavier than white, bright colours look heavier than dull colours.*
- ❖ *Complex figures appear heavier than simple ones.*

Direction works like this:
- ❖ *The shape of objects creates an axis and thus direction.*
- ❖ *Direction is also influenced by the arrangement of surrounding objects.*

Balance and symmetry are affected not only by the arrangement of objects, but also by the division of space across the whole map sheet. Unequal divisions of space create a more interesting layout, although an overall equilibrium should be maintained.

If you are making a series of thematic maps, make the layout of the legend elements the same for each map in the set. Maps of the same area will have a map border the same size and shape. However, be sure to consider that one thematic map may have more symbols than another when you choose the placement and size of the legend box.

When satisfied with the layout, use a straightedge to draw lines for the border of the map and the legend boxes. Cut the paper to size.

12.3 DRAFTING (DRAWING) THE FINAL MAP

This section is about how to hand-draw or *draft* maps. This section will describe special drafting instruments and papers and techniques to use to achieve a professional product. Almost anyone can produce professional-looking maps by using the right tools and materials, and by following the right order of steps.

12.3.1 DRAFTING MATERIALS

Most manual drafting materials are available in major centres around the world. Even if you are located in a small town or a remote area, you should be able to order materials and equipment from a drafting supply store in your capital city.

Drawing Paper and Plastic Films

Widely available kinds of drawing paper include *regular paper*, *graph paper*, *drafting paper*, *tracing paper*, and *transparent plastic film* (acetate, polyester) and *mylar* (described below). Because ink is in many ways better than pencil lead for drawing final maps, all proper drafting papers and films are finished to take ink evenly. For a neat-looking map, the inked lines must be sharp and of regular width—not fuzzy or irregular. Regular paper and tracing paper may absorb ink, producing a fuzzy line, or smudge easily, but these papers can be used if that is all that is available.

Tracing papers are translucent and are excellent for the purpose of re-categorizing and transferring information—particularly when a photocopier is not easily accessible, or when you need to be selective. Using tracing paper, you can draw the information in categories or layers, one map sheet for each layer of information. Or you can combine information from two maps to summarize the information.

However, tracing papers can stretch or shrink with changes in temperature and humidity, and are therefore difficult to use for overlays—the scale keeps changing— except under consistent conditions. Inexpensive tracing papers, the only ones available in some places, warp from even the dampness of a sweaty hand. Stretching, shrinking, and warping will obviously affect the accuracy of the map. Therefore, it is good practice at all times to rest your hand not on the paper, but on a clean, dry cloth.

Two specialty drafting papers are vellum and mylar. **Vellum** is translucent like tracing paper but takes ink better. Corrections are not easily made, however, because erasing pencil or ink with an eraser (rubber) can damage the paper. **Mylar** is a whitish plastic film 'paper' that is textured ('matte' or 'frosted') on one or both sides to take ink (or pencil) well. Errors are relatively easy to correct, with either a special ink eraser or a slightly moistened white eraser (an **erasing shield**, a thin steel plate with different shapes of holes in it, is handy), or with a gentle scraping action using a very sharp knife (be careful to remove just the ink). Mylar doesn't shrink or stretch, but use a cloth or piece of paper to rest your hand on anyway, because oils from your skin can cause the ink to bead on the mylar. Mylar is an excellent material for

Drawing a map is easier if you have a large table to put it on.

mapping; however, it may be available only in large cities, and it is the most expensive kind of drafting paper.

For specific purposes, maps can also be drawn on sheets of **transparent plastic film** (acetate or polyester). Transparent sheets are used to make overlays, so that the user can compare one map with another visible underneath. Small-scale maps can be drawn or photocopied onto transparent A4 size sheets and shown with overhead projectors. Transparent sheets are also used for multi-colour maps, where each sheet (mylar sheets could be included too) will be printed as a different colour or tint (but note that the artwork on the sheets themselves should be black).

Use colour felt pens (markers) designed especially for use on plastic film or those designed to 'write on anything' (ink from waterproof or 'permanent' felt pens can usually be removed from the sheets if necessary with a scrap of cloth moistened with some isopropanol (rubbing alcohol)), but most kinds of drafting ink will bead on a smooth surface. Stick-on symbols and patterns work well, but rub-on ones don't.

Pencils and Pens

Pencils are used for the working stage of the final map because it is easy to make corrections. *Mechanical pencils* are easier to keep sharp and are therefore better for drawing clean and regular lines and for fine lettering. Leads are available in different softnesses. They are coded 'B' for soft and 'H' for hard, and also numbered—the lower numbers are closer to 'HB,' an average hardness. A code of '2B' indicates a relatively soft lead and '2H' a relatively hard one. A soft lead shows darker on the paper but has a tendency to smudge. A hard lead shows up lighter on the paper; a 3H or 2H lead is often used for temporary (construction) lines. However, a pencil that is too hard can damage the paper by cutting into it, especially if you have a soft or irregular surface beneath it. Although coloured leads and pencils are also available, they are normally used only for fieldwork or for hand-coloured maps for display.

Technical pens are designed to facilitate drawing lines of even widths. They consist of a tubular point, which is precisely manufactured to a specific size, and an ink reservoir in a holder. A different pen (or nib) is used for each line width. You may need a set of three to six nib sizes, typically between 0.3 and 1.0 mm. Be careful to always replace the cap on the pen when it is not in use—if the ink dries out, it clogs the tip (especially the smaller sizes) and stops the flow of ink. Occasionally you will have to partially disassemble the pen mechanism and clean it with warm water or special cleaning fluid (take care not to lose any small parts). You could also, less conveniently, use a dip pen with the appropriate style of nibs.

Always use standard **black drawing ink** for drafting a final original map that is to be duplicated (unless it's by full-colour photocopier). Drawing ink is chemically based and thus retains its blackness over time (nevertheless, don't leave your map in the bright sun for week after week) and is not affected by humidity, so it won't smudge like some other inks will, but it's still important to keep your map dry. Drawing ink is available in bottles, cartridges, or tubes.

Drafting Instruments

Drafting instruments are tools used to make measurements and to guide pens. The most important tools for drawing are a **metric ruler** for measuring and for drawing straight lines (to also have a separate **straightedge** is better); a **T-square** and a **large**

TOOLS AND MATERIALS FOR HAND-DRAFTING A MAP

☞ *Tracing paper (for better maps, use drafting paper—such as vellum or mylar)*

☞ *Pencils*

☞ *Eraser*

☞ *Sharp, small-bladed knife (such as X-acto brand)*

☞ *Technical pen(s)*

☞ *Ruler (and straightedge if possible)*

☞ *T-square (and large triangle if possible)*

☞ *Protractor (360° if possible)*

☞ *Templates or adhesive patterns and lettering*

---

triangle for drawing perpendicular lines; and a **protractor**, for measuring angles, such as compass bearings. Because you are probably using angles from 0° to 360° from compass bearings in the field, it is easier to use a round 360° protractor rather than the more common half-circle (180°) one. In addition, you will find it useful to have **flexible curves** or a set of **French curves** for drawing curves with changing radii, and a set of **templates** for drawing various shapes—such as circles, squares, triangles, and arrows—and for letters. However, a steady, practiced hand can also produce very good results.

Note that tools made for use with ink have a beveled (undercut) edge to stop ink from creeping underneath. If you need to, you can provide similar protection by applying strips of tape to the bottom of the tool to raise it slightly above the drawing surface. The outside edge of the tape should be about 2 mm from the edge of the tool.

Working Surfaces and Tables

Drafting is best done on a smooth surface. For your comfort and ease of working, **drafting tables** are made so that the table top can be adjusted to a comfortable working angle. This angle helps prevent back strain. However, you can also use a regular table instead. It is most important that the working surface is flat and smooth—be sure to protect it, such as by not cutting directly on it with a knife. Glass makes a good surface. If you have a flat surface that isn't quite smooth, you can buy a rubber drafting surface (thin mat) at specialty stores that sell technical instruments. A **light table**, a glass-topped box (often with legs), has electric lights (usually fluorescent tubes) inside it that can shine through several layers of drafting paper or mylar to reveal what is on underlying layers. A light table is especially useful for tracing and is also fine for general drafting. You could easily make one yourself if you have experience with carpentry and electrical wiring.

12.3.2 DRAFTING TECHNIQUE

How good your map looks depends on your drafting materials and your technique. For people with reasonably good hand-eye coordination, technique is simple. You need a steady hand and concentration, and you need to follow the right order of drawing steps. Consider using the suggested orderly drafting process on the next page. In this case, it is for the process of tracing the information from a field map (topographic base

AN ORDERLY DRAFTING PROCESS

1. **Make four registration marks** (+) *on the field map or base map. Fix the map to the working surface so that it won't move. Masking tape, placed outside the drawing area, is usually used. Drawing pins or needles can also work, but they will damage the surface beneath the map.*

2. **Draw the outer frame or border** *on the tracing paper (or other transparent drafting paper). Use a right-angle ruler, T-square, large triangle, or a protractor to make sure that the frame is rectangular. (Each corner should be a 90° angle.) Frames for all the thematic maps in a series should be drawn the same size.*

3. **Draw the inner frames** *that will surround the map itself, the legend, and the inset map. Make sure that these boxes too are drawn exactly rectangular, with their edges parallel to the outer frame.*

4. **Precisely line up the tracing paper on top of the field map (base map)** *so that the sides of the outer frame on the tracing paper are parallel to the grid north lines on the field map, and so that the map information is centred. Fix the tracing paper in place using masking tape. Make sure that the tracing paper is flat and not warped.*

5. **Trace the registration marks** *from the base map onto the tracing paper. Mark the latitude and longitude lines (or UTM lines, or both) in the margin only—don't draw the grid lines across the map.*

6. **Draw the north arrow.**

7. **Draw the details of the map, starting with any straight lines and smooth curves (major rivers and roads),** *because they are the most difficult to correct. In general, work from the centre of the map to the outside.*

8. **Then draw the rest of the rivers, boundaries, and roads.** *Start with the longest first. Note that you probably won't want or need to draw the contours lines from a topographic map.*

9. **Next, draw the main point features** *(or their centre points if you are using stick-on symbols); they may be villages or houses or even gardens, depending on the scale of the map. If you are tracing from a topographic map, mark each mountain peak with a triangle and write the elevation (as written on the topographic map) nearby.*

10. **Add the lettering.** *If you are lettering freehand, don't bother to trace the lettering of the base map—your own style will generally be more consistent, and you can adapt the size as needed. If you are using labels printed out using a computer or typewriter, wait until step 13 before applying them.*

11. **Add the area symbols, if you are drawing them freehand.** *(If you are using adhesive patterns, then wait until step 13 before applying them.) Doing the area symbols last allows you to make sure that you won't obscure other information with them.*

12. **Ink the map in.** *It is often easiest to draw all the features of the same type at the same time, so that you don't have to change pen sizes so often. Be sure that the ink has dried before you put a tool or cloth on it! (Draw all the lines that run roughly in one direction before switching to those that are roughly perpendicular to the first set.)*

13. **Finally, apply any adhesive-backed materials,** *commonly used for area symbols— or for certain point symbols. If you are using symbols or labels that are printed on white paper, be sure to trim off the excess paper first, leaving a border of about 1–2 mm, or you may obscure too much of the underlying information.*

map or plotted survey) onto tracing paper (or other transparent drafting paper), where the traced map will be your final map.

Here are a few additional reminders before you start drawing:

❖ *Keep a record of the line weights (pen sizes), and any codes for commercial patterns and letter sizes you are using in case you need to make changes later, or to match other maps in the series.*

❖ *Do step 1 to step 11 in pencil, in case you find that you need to make changes, before you use any ink on the map.*

❖ *Pencil lightly, and never let the pencil dig into the paper (pull, don't push, your pencil).*

❖ *Be gentle using an eraser, because it can damage the paper and affect the accuracy of the map.*

❖ *Some kinds of tracing paper warp just from the moisture of your hand. If you rest your hand on the map while drawing, put a clean, dry cloth under your hand to prevent the paper from soaking up humidity or oils from your hand, and to prevent smudging.*

❖ *When you are not working on the map, cover it with a cloth (or a large sheet of paper or plastic) to protect it from dust and marking.*

12.4   CHECKING AND VALIDATING THE MAPS

The community must have the opportunity to check the maps before the maps are finalized. This part of the mapping process is very important. Community validation is a test of how well the information presented on the map represents reality and the perception of the community. We cannot say that the maps are correct until after they have been checked by the people in the village.

Many communities have had the experience of sharing information with outside researchers or government officials but never seeing any of the information come back to them to be checked—or for their own use. Bringing the maps back to the community for villagers to check assures them that the maps are accurate and that the mappers heard and understood their information and concerns. Villagers will feel more inclined to stay involved, to continue to add to the maps, and to actually use the maps.

When should community members review the maps? They need to check them after the final map format has been designed and the information has been transferred from the field maps and notes to the final, or semi-final map. This way the villagers can see exactly what the final maps will look like. These draft maps could be drawn only in pencil—that way it will be easier to make changes than if they have been inked in and stick-on patterns applied. (Note that it is possible to blueprint or photocopy a map drawn in HB pencil, so that any comments and changes can be sketched in on the copies. The changes can later be neatly drafted onto the originals.)

Checking at this stage verifies both that the information was understood and recorded correctly in the beginning and that the drafters transferred the information correctly. The community members are then assured that the maps accurately represent the information that they gave you.

Maps are best validated in group meetings, by as many of the original contributors as are interested in taking part. If new potential contributors show up at this late stage, you will need to work tactfully with the group to decide whether the value of any new information outweighs the potential delay in map production.

Have a member of the community mapping team present the maps at a village meeting. The purpose is for all the villagers to understand the maps enough to be able check the validity of the maps from their own perspective. It also may be an opportunity for initial discussions about some of the things that the maps highlight. Make sure that the villagers understand the maps. If necessary, divide into small groups and have a member of the community mapping team explain the maps to each group and answer any questions. Let the villagers know that it is okay to draw corrections directly on these draft maps (or copies of them).

People may need more time than a two-hour meeting to study the maps and make corrections or additions. If so, after the village meeting, you could post the draft maps in the community hall or a similar public place and allow people to study the maps at their convenience. Encourage people to draw or write corrections on the maps. Post extra blank paper so that people can make suggestions or draw sketches. Instruct them to write their name or initials along with the corrections or suggestions so that the mapping team can question the author later if what they wrote is unclear.

COMPUTER-ASSISTED MAPPING AND GEOGRAPHIC INFORMATION SYSTEMS (GIS)

Another way to produce maps is by using a computer. In the long term, the use of a computer offers several advantages, giving you the ability to produce maps quickly, change and update the maps more easily, make analyses and calculations, and produce professional looking map designs with colour. However, computer mapping has the disadvantage of taking a long time to set up and learn, and of being much more costly. In the short-term, and for small mapping projects, it is faster to produce maps by hand. In any case, the best way to learn about the principles of mapping and cartography is to first get experience with making maps by hand. Of course, another disadvantage of

using computer mapping is the tendency for it to become a tool that is used only by one or two people in the community. To use the GIS effectively in participatory mapping, the majority of community members must understand enough about the GIS to be able to realistically decide what they want to put into it and what they want to get out of it.

13 WHERE TO GO FROM HERE— USING THE MAPS

What you will learn in this section:
> ➤ *Five basic purposes for which maps can be used to benefit local people*
> ➤ *Examples of what some communities have done with maps*

13.1 MAPS ARE A TOOL

Maps are a powerful communications tool. Because they are visual and graphic, maps are an effective tool for community members to communicate about land issues— amongst themselves or with outsiders such as government officials or industry stakeholders. Maps can help the community to articulate its history and its culture, its problems and its plans.

Every community needs a unique strategy to use the maps that they've made, depending on the culture, the land issues, and the political and legal contexts. This section will give you some ideas to think about for your own community by presenting various ways that rural communities around the world have used maps. They have used maps of their community lands as a tool to

❖ *Document and preserve local/traditional knowledge about the land*
❖ *Plan and manage community lands*
❖ *Raise community awareness about local land issues and motivate communities to address them.*
❖ *Increase local capacity to communicate and work with external agencies*
❖ *Gain recognition of customary land rights.*

Mapping community lands is often an ongoing activity. A community creates, updates, and over time builds a **map database** (a series of maps with background information) that can be used for any or all of the above purposes. Very often, recognition of customary land tenure is the ultimate goal. But, depending on the case, a community usually needs to use the maps to address some of the other issues first, such as raising community awareness and achieving unity, or making a community land-use plan that can be presented to government.

Some communities have a well-defined strategy before beginning to make maps, and some do not. If maps are designed for a particular strategy, then the mapping process is more efficient, so that villagers collect only enough information to present their case. But even map information that is not directly relevant to a particular strategy will be useful at least for a record, and collecting it is not necessarily a waste of time. As explained in chapter 4 in connection with planning a mapping project, a community must envision a purpose (or several) for the maps in order to be motivated to make the maps. And the purpose provides some guidelines for deciding what to map and how to map.

MAPPING DAYAK LAND USE IN WEST KALIMANTAN

In West Kalimantan, Indonesia, an organization called PPSDAK trains Dayak villagers to make maps of their traditional land use. PPSDAK has refined the methodology over a few years. It goes like this:

1. **First contact:** *A village contacts PPSDAK and requests assistance. A facilitator travels to the village to get to know the people and discusses with villagers what is involved in mapping.*

2. **Letter of request:** *A letter is sent by the villagers, requesting training and technical assistance and agreeing to contribute food and volunteer time.*

3. **Technical reconnaissance visit:** *PPSDAK staff travel to the village to make the technical preparations. They take a GPS coordinate at the centre of the village and estimate the village area. With this information they can estimate how long it will take to do the mapping, they can order and enlarge the topographic map to the appropriate scale (usually 1:10,000 or 1:20,000), and prepare the necessary tools.*

4. **Community meeting:** *The village hosts a meeting with people from surrounding villages to discuss traditional laws, especially with respect to the boundary.*

5. **Training:** *PPSDAK staff spend three days training villagers in the central and surrounding villages in how to read topographic maps and how to use a compass and the GPS.*

6. **Technical mapping process:** *Villagers build a 3-dimensional model of the land using the topographic map. The model is used for village discussions about the location of land use, forest products, and cultural sites—information to be drawn on maps later. Next the villagers survey the settlement using compass and metre tape. Then they survey the customary (adat) boundary using a GPS receiver. During the evenings the villagers gather and draw from memory on the maps.*

7. **Drafting the maps:** *When the field surveys are complete, two community mappers from the village accompany staff to the offices of PPSDAK. There they process the survey data and compile the results with the sketch maps. Then they draft a series of six maps that show the village culture and land use.*

8. **Verification:** *The community mappers take the maps back to the village to be verified by the whole community and signed by the village leaders and elders. The original final maps are kept in the village and copies are kept at the PPSDAK office.*

This process has been evaluated by villagers, NGO peers, and the international agencies that fund the program. The proof of success is in the 80 villages who have already made a complete map series of their land, and in the many requests that continue to come from additional villages.

The maps were well received by government officials when the villagers presented their maps in a large seminar organized by PPSDAK. Officials from the provincial land-use planning department were impressed with the technical quality of the maps and surprised at the villagers' ability to make them. The result is that officials have promised to consult with villagers when planning future land-use developments. The village maps will be at the centre of these consultations.

—printed with permission of PPSDAK

In fact, some communities are motivated to make maps by just an intuitive understanding that the maps are powerful information to be used for the general purpose of protecting the lands. Through the process of making the maps, a strategy for using the maps gradually emerges. But, a word of caution: Without a defined purpose or strategy, be careful that the community is not motivated by false expectations that the maps alone will miraculously resolve the land issue that they face. The maps are only a tool—the majority of the work comes after the maps are produced and it is time to present the maps in ways that benefit the whole community.

13.2 MAPS FOR RECORDING AND PROTECTING LOCAL KNOWLEDGE ABOUT THE LAND

Many communities have used maps to record the knowledge of the elders about the land. For many cultures, their history, their traditions, their pride, and their identity are written on the landscape. When the traditional knowledge about the land dies, so does the culture. Around the world, a great deal of local knowledge has been lost because village elders have died without passing on their knowledge.

Any map—a sketch map or a topographic map—is useful for recording traditional knowledge. The important thing is that when the elders talk about a place you can find it on the map and later, using the map, find it on the land. Only then can you consider the map to be a permanent record. In Canada, aboriginal communities use the term 'Traditional Land Use Study.' Such a study always includes maps and may also include photographs, written stories, archaeological artifacts, and plant collections. Taken together, this 'data' documents local knowledge.

Of course, a map is only a tool, albeit a very useful one, for recording and conveying a selected portion of a people's knowledge—it can hold only a small portion of the richness of the culture and tradition it represents. Maps of local knowledge are usually drawn as thematic maps. Some types of traditional knowledge commonly recorded on maps include

- ❖ *Migration: Where the local people of the village came from*
- ❖ *Settlement: Where the local people of the village settled*
- ❖ *Traditional land use: Where people farm, hunt, fish, or pasture animals*
- ❖ *Territorial boundaries: Where the boundaries are between one village and another, or between people who speak the local language and those who speak another one*
- ❖ *Traditional resource ownership and tenure systems: How people divide the land, and who has the rights to farm or hunt in certain areas*
- ❖ *Cultural and sacred sites: Where the gravesites are, and where to find the sites with important stories*
- ❖ *Local ecological knowledge: Where the wild animals feed and breed, or where certain plants grow, or where the soil is good for farming, and when the weather is suitable for planting*

Why is it important to make a record? The elders used to share this knowledge of the land with the young people as they worked together in the fields and in the forest. Knowledge of the land has traditionally been passed from the older generation to the younger generation through stories, songs, and daily work and living. These days, maybe the young people spend much of their time at school and don't have the opportunity to learn how to find or identify useful plants to make traditional crafts and medicines, or where the traditional boundary is. Knowing about the forest and the land and their

THE AKHA OF NORTHERN THAILAND

The Akha people came to Thailand many decades ago. They settled in Northern Thailand in the mountain areas that the Thai people were not using. They chose their new village sites carefully—usually high on ridge-tops.

As they have done for generations, they maintain a ring of protected forest around the village for a distance of about 1 km. Outside that ring they maintain another ring of forest that they use for wood to build their houses, and outside that ring they establish rice fields. They mark many ceremonial and sacred places with special ornaments made of wood or bamboo. All of these places have names.

In 1994, some Akha college students learned to survey and draw maps. They worked with the Akha elders to make two maps for each village, one of the village itself and one of the whole territory, extending beyond the rice fields to the boundary. They sur-veyed the village and marked the names of the people in each household and drew in the ceremonial places. To make the territory map they used a topographic map as a base map, and used compasses and the triangulation method to draw the outer boundaries, the rice fields, the fruit trees, the vegetable crops, and the forest rings.

Although many Akha children now go to Thai schools, when they look at the maps of their home place—showing the sacred places, the fields where they help their parents, and the places where they listen to Akha ceremonies and songs— they feel proud to be Akha.

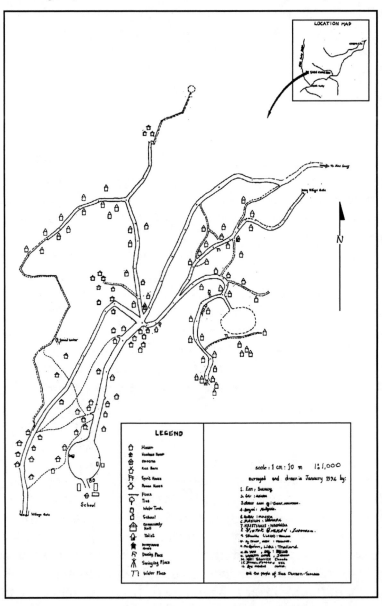

own history is important for children to be certain and proud of who they are and where they come from. Children often believe that the school knowledge is more important than the traditional knowledge, because they expect that it will give them the opportunity for jobs and to be part of the larger society. Indeed it may, but children will be richer in knowledge and self-worth if they appreciate and learn both sets of knowledge.

Maps are one means to make a permanent record of the community lands, one that can also be used to teach children about the land. Having a permanent record of the land on a map—like they see in school for other places—will make the children feel proud of the place that they come from.

Maps are a record that can help to protect local knowledge from being lost from generation to generation. Some people might say that maps aren't necessary for this purpose because local people already have customary ways to pass knowledge from generation to generation. Of course, elders should continue to use the teaching ways customarily employed by the traditional owners of the knowledge. These traditional methods are the best way to keep the knowledge system alive. However, if you are concerned that, for any reason, not all of the knowledge is being passed down, then maps are one good way to make a record of it. But respect the traditional laws. If there is some knowledge that is secret or is not allowed to be written, then find other ways to ensure that that knowledge is passed down.

One way to think about it is that elders have the responsibility to their ancestors to express their knowledge to the next generation, and youth must accept the responsibility to guard that knowledge from exploitation and use the knowledge only in accordance with traditional laws.

Indigenous people's ownership and custody of their indigenous knowledge is collective, permanent and inalienable, as prescribed by the customs, rules, and practices of each people. Local people own the maps and the knowledge on them. This ownership can be affirmed by writing a copyright statement on every map. However, this precaution alone is not sufficient to protect the community's intellectual and cultural property rights. For maps to be used effectively for maintaining and guarding traditional knowledge, the community must design a strategy to control access to the information on the maps. This process often entails designing some maps to show information that is used within the community only, and other maps to show information that may be shared outside the community.

13.3 MAPS FOR RAISING COMMUNITY AWARENESS ABOUT LOCAL LAND ISSUES AND MOTIVATING COMMUNITY MEMBERS TO ADDRESS THEM

Maps can be used to raise awareness about local land issues. Maps provide a visual medium to stimulate people to think about the environment, to talk together, to understand their interdependence with their neighbours, and to understand the environmental consequences of their own actions, or the actions of other parties, such as a logging or mining company. These maps can be used to help motivate people to collectively take responsibility for caring for their environment.

Land issues are becoming more and more complex. In many communities, there was once enough forest and clean water for the neighbouring villages to share without conflict. But perhaps now a village upriver has started planting cash crops and they use chemicals to fight the insects. These chemicals might contaminate the water of the communities downstream. Maps can be used to show villagers how their actions affect their own environment, and how what they do in their local environment affects the environments—and people—of other villages. Colourful maps, or three-dimensional maps (for example, of cardboard and plaster) can be used to clearly show everybody in the community how your local environment is connected to other villages—or a plantation development upriver.

A meeting of the community, using maps to discuss land-use issues.

Some environmental problems are a result of relationships between people within the community. Maps can show the distribution of lands among families and clans, and how this distribution is related to effects on the environment. Maybe a family does not have enough land to rotate its crops and so the family has to make new gardens in the same place year after year, thereby depleting the soil. On the other hand, perhaps a family with a lot of money buys and uses pesticides, which affects the land of the neighbouring family. In this way, maps help to illustrate the relationship between local environmental concerns and local land-use decisions.

Of course, maps can also help to enlighten villagers about the potential impacts of proposed industrial developments. Lay a map of a proposed logging plan over a map of

THE SAM–MUN WATERSHED

The Sam–Mun watershed in northern Thailand is a critical source of water for 60 villages. This water source is threatened by deforestation and by the use of agricultural chemicals. The Royal Forestry Department of Thailand helped community organizers and villagers construct three-dimensional models of the watershed by using topographic maps and aerial photographs. These models were used in community discussions and helped villagers to understand the upstream-downstream links to each other and the implications of deforestation. Villagers then came up with their own suggestions for resource regulations.

—from Limchoowong, 1992

the community land and see exactly what land is intended to be logged. You can now answer questions such as *Is it owned by one family, or is it communally used land? Is it land that the community would farm? Or is it land that is protected by traditional law? Is it an area with high value non-timber forest products? Is it a sacred area?*

Or if resource extraction has already occurred, maps can be used to provide a picture of the environmental impacts. A map can be drawn to simply show the location of environmental problems such as soil erosion, land slides, road washouts, chemical spills, damage to fish habitat, or damage to bird nesting sites. A map of environmental impacts could also show how these problems affect people in the community. Show the place where villagers get their drinking water in relation to environmental problems upriver. Or make a map that emphasizes the impacts on the habitat of a forest animal species that is particularly important as a local food source. Maps are a good tool for drawing relationships between environmental problems and the effects on the local way of life.

As visual tools, maps facilitate discussion about these questions and issues. Maps attract people to discuss land issues and seek solutions. Not only that, but the process of making maps can help to bring people together—to put their knowledge together to make a concrete product. Once the mapping project has been initiated, no one wants to be left out of the picture. And, in the end, no one person owns all the knowledge on the map produced. It is only possible as a collective effort.

People in the community can see their common history on a map. Seeing a picture of their place and their history together helps to strengthen the community. Perhaps for the first time people see on one picture the land that they share, and realize that they also share responsibility for protecting it for their grandchildren and generations to come.

13.4 MAPS FOR PLANNING AND MANAGING COMMUNITY LANDS

There are more people on this planet, competing for land and resources, than ever before. Many local communities find that they have to struggle to maintain lands that were passed down to them by their ancestors. The members of each community must think about their future before an outsider plans their future for them. Maps are a helpful tool for participatory planning and management of community lands.

Planning is a process of envisioning the future, then analyzing options, making decisions and taking actions to make that vision happen. Particularly for land-use planning, maps are a visual tool that can help each community member to see how various options might affect their own future. Villagers can draw maps to show several possibilities of what they want the land to look like in one year, two years, five years, or twenty years. Maps are very helpful to do this kind of planning, because they are pictures that everyone in the community can understand.

In the past, communities didn't need to make formal plans, because they followed the traditional laws and wisdom of the elders. But now, as a result of a variety of commonplace scenarios, things are changing. Industrial developments may be exploiting resources nearby or within the community's lands, in the process polluting river water, thereby reducing fish populations, and cutting forested areas, thereby destroying animal habitats and populations. Because of these developments, hunting and fishing might not be as easy as before. There might not be enough community-controlled land

THE GITKSAN: THE LAX' SKIIK AND
THE STRATEGIC WATERSHED ASSESSMENT TEAM

The Lax' Skiik (the Eagle Clan) of the Gitksan people live on their traditional territory in northern British Columbia, Canada. There have been many uninvited activities on their land for over a century. The Lax' Skiik want to develop a sustainable and environmentally sound approach to living on their tribal territories. To do this they need to know what resources they have and what their own needs are so that they can continue to survive for generations to come.

Together with the Gitksan's Strategic Watershed Assessment Team, the Lax' Skiik have developed a model for wildlife inventory and habitat assessment. The Lax' Skiik team runs transects—with compass or GPS—through different forest types, on varying slopes, and at varying elevations. This way they can identify and map what species of animals live in these areas and the types of habitats that each species needs to survive in all four seasons, as well as the animals' travel routes from one habitat to another. While doing the transects, if the mapping team sees important medicinal plants or culturally modified trees or an old hunting camp, these features are also recorded and mapped.

This method is a very time consuming. But it fulfills several important objectives: for example, to allow Lax' Skiik people to learn new skills out on the land and to help the Lax' Skiik make decisions that care for the land according to the Laws of the Creator—and according to ecological science. The Lax' Skiik have been working on this project for two years, and they expect it to take three more years for them to finish this work on their territory.

With this inventory and their maps of medicinal plants, forest foods, wildlife, water, and historical places, the Lax' Skiik want to enhance cultural education and tourism. Youth groups, parents, and anyone interested will be able to learn how the Lax' Skiik tribe uses the land while respecting Cont'd on p. 183

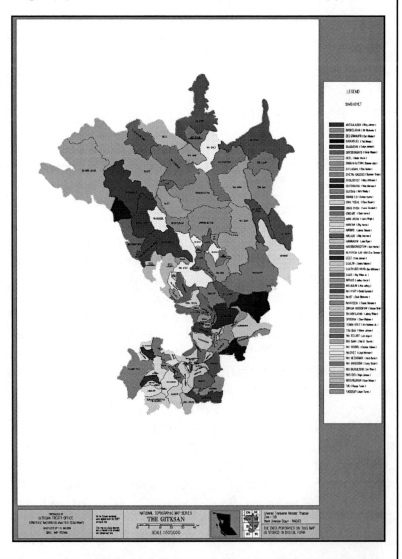

anymore to rotate the farmed fields so that there is a fallow time during which the soil can regenerate.

The impacts are not only environmental, but social as well. Young families want more cash to send their children to school, and so they prefer to work for a logging company rather than farm the land. The people's way of life may be undergoing great changes. As developments encroach on community lands or the village population grows, the distribution of land and land use becomes more complicated. If the available land is smaller than it used to be, villagers have to choose more carefully where to make their gardens. The village must decide how much land is necessary for food crops and how much is available for cash crops. The villagers now need to make a land-use plan.

Maps can help villagers to make land-use plans in which they record which areas they have chosen as being best suited for food crops, for cash crops, for hunting, for building houses, etc. The maps provide the whole picture, so that the community can discuss together any topics about how the land is used; for example, for developing cash crops, regulating hunting or fishing, building a road, developing a tourist area, or building a small dam.

Maps are a good tool for planning land-use activities, if they can show
- ❖ *What resources exist and where*
- ❖ *What areas have potential for what kind of land use*
- ❖ *What areas need to be protected for cultural (customary or* adat*) or environmental reasons*

Land management means regulating and monitoring land-use activities. When communities manage their own lands, they regulate their own activities. Some examples of regulations are
- ❖ *How many hectares of forest a family may cut and farm each year*
- ❖ *Where people may cut forest for gardens and where they may not*
- ❖ *How many wild pigs may be killed by village hunters in one year or season*
- ❖ *Where hunters may hunt*

Indigenous communities usually have traditional laws that apply to land-use activities. But if there are new pressures on the land (more people on less land), then new regulations may be required to make sure that river water stays clean, and that fertile soil and animal and plant populations are not depleted. These new regulations should be decided by the whole community. Maps are a useful tool to make land issues visible and to help the community to discuss and decide whether new regulations are required.

Maps also provide a framework for collecting data to monitor changes on the land. Instead of writing notes, or simply telling your friend, you can draw it on the map. Make a map showing where a logging company logged last year and this year, and where there is excessive erosion. Or make a map showing which family is farming

---

the Laws of the Creator for all living and non-living things. The teaching will take place in longhouses situated in three different ecosystems: river, mid-elevation, and alpine.

The maps will eventually be used to develop a whole landscape plan for Lax' Skiik territory. This process will involve analyzing the inventory data and maps to identify ecosystems and habitats that need to be protected in different ways, as well as areas that have local economic potential.

—written by Art Loring

THE TOLINDU

The people called the ToLindu live in three communities on the shores of Lake Lindu in the mountains of Central Sulawesi. They care for the fish populations in the lake and each family has a designated area for fishing. They grow rice on the flats near the lake and they hunt animals and gather plants in the surrounding forest. The government designated part of their traditional lands as a national park. At the same time, plans were being made to build a hydroelectric dam at the outlet of the lake. Both of these government land developments would have serious impacts on the Tolindu way of life.

The three communities chose to work together to make several maps. One map shows their traditional land-use system: where they grow coffee, fruits, and rice; where they have grazing areas; where they hunt animals; and where they gather medicinal plants. Another map shows the boundaries of their communities and indicates where traditional laws affect the use of resources. For example, it shows forest areas that are protected and other areas that may be farmed.

With these maps, the ToLindu expressed to government decision-makers their concerns about the impacts of the developments planned for their land. They demonstrated that the ToLindu people know more about the flora and fauna of the area than anyone else does. They also showed that they are capable of making technical maps that display their traditional knowledge. By expressing their knowledge in a way that government officials could understand, the ToLindu demonstrated that they were capable of entering into technical discussions and contributing to decisions that would affect them, and could join in partnership with the government.

The Tolindu villagers used the maps to discuss the zonation of Lindu National Park with the government's parks managers, with the result that the ToLindu territory was designated as an enclave in which villagers may continue their traditional way of life.

—written and photographed by Alix Flavelle, with
the permission of the ToLindu village council

where, and who is starting cash crops. A map is an accessible record of such information. By comparing maps from before and after a certain event or proposed change, villagers can visually see changes and discuss them. With maps you can illustrate answers to questions such as the following:

❖ *Is there less forest than a few years ago?* (If you have a map made three or five years ago, you can compare it to one that you make now. If you don't, you can make a map of the past from memory, but it will be less accurate.)

❖ *Are hunters coming onto community land without permission?* (You can show on the map where are they hunting, compared to where they used to hunt, if you know.)

❖ *Is logging making the rivers more muddy than two years ago?* (Show on the map where the logging happened in the last two years.)

13.5   MAPS FOR INCREASING LOCAL CAPACITY TO COMMUNICATE WITH EXTERNAL AGENCIES

To resolve land-use conflicts and issues involving external threats requires communication with external agencies. Often, a large part of the problem is that the government department or the project developer does not understand the significance of the land for the local community. Land-development decisions are often made in far-away offices, where perhaps the small-scale maps being used don't even show that a village exists there. Even if local people try to explain their concerns about the development project, officials do not understand, because they don't understand the local economy and way of life.

Maps are visual pictures and thus provide a common language to help local people communicate with government departments and project developers. Government mapping styles don't capture the importance of the land uses of the local community. Indigenous subsistence land uses are often invisible to government agencies. Locally made maps can support indigenous peoples' insistence that the land is already fully occupied and used. Villagers can easily use maps to explain their situation, to show the importance of the land to their local economy, and to show their history on the land. Then, after looking at the maps, outsiders can ask more meaningful questions and dialogue can begin.

Sometimes government departments require official-looking communications in order to know how to respond. In this way, maps can be used to make formal applications for community forest reserves, or formal application for land title, or for negotiating compensation for land or resources that have been taken from the villagers.

Community maps are just a tool to facilitate communication between local people and outside agencies. Just as important are the activities behind making the maps—field surveying, expressing local knowledge, telling stories, holding community meetings, using the maps to discuss land issues internally. All of these activities help to prepare the people to voice their concerns about the land with greater clarity and courage.

13.6   MAPS FOR ESTABLISHING LAND RIGHTS

Maps are an essential tool for establishing aboriginal rights to land and resources. Maps are effective for clarifying land boundaries and for depicting historical land use, occupancy, and traditional ownership of the land. Strategies for establishing land rights fall into two general categories: legal (through negotiation or litigation) and political (through lobbying and policy reform). Both rely on proof of historical precedence on the land.

In Canada, maps have been used as evidence in the courts for aboriginal land claims. For a map to be admissible evidence in court, there must be 'witnesses' from the community who can appear in court to declare the validity of the map. This means that during the mapping process it is important to keep good records of dates and names of the people giving each piece of information about particular sites or areas or boundaries. Surprisingly, the quantitative accuracy of the map matters much less than the *consistency* of the data between what is said verbally or in written form and on the map—hence the importance of finding reliable sources of information and of recording it accurately.

In most cases, mapping the outer boundaries of a territory is probably the most important first step. Most legal land-title documents contain a written description of a boundary as well as the surveyed map. Likewise, the boundary of the traditional territory should be described in writing with reference to landmarks as well as drawn on the map. Written accounts of historical events or agreements regarding the boundary also help to validate how it is drawn on the map.

In Canada, there have been two basic types of legal argument for land rights. One type is based primarily on historical and present land use and occupancy. The maps used to present this kind of argument would show areas that are historically and currently used for hunting, fishing, and plant gathering activities. The other argument is based primarily on oral history—demonstrating that traditional ownership rights are recognized and practised from some time in the distant past up until today. Maps drawn to support this kind of claim would show place names, sites of historical events, and the boundaries and land use associated with family lineages.

Legal strategies are often used in combination with negotiations. Negotiations often entail a quantitative analysis of maps, calculating how much land area is at stake, as well as the quantity and value of different resources. This approach obviously calls for reasonably accurate scale maps, as well as resource inventories.

Maps can also be used in a political approach to establishing land rights. There are a variety of political tools to gain the attention of politicians, government officials, and the public. Once you have their attention, maps can be an effective tool for conveying the message that the indigenous community exists, that it has relied on and stewarded its ancestral lands and resources for generations, and that it therefore has definable and inalienable rights.

Regardless of the strategy, the process of establishing land rights can take decades. Obviously, maps represent just one tool in the process. The maps will come under intense scrutiny. The key to the successful use of the maps is to establish within the community a good communication and decision-making process for creating truly representative maps, presenting them, and defending them.

APPENDIX A

COMPASS SURVEY NOTES

Name of village _____   Survey line _____

Name of chief surveyor _____   Date _____

STATION #	FRONT BEARING	BACK BEARING	SLOPE DISTANCE	SLOPE	HORIZONTAL DISTANCE	NOTES	

GPS DATA

Name of village _____   Survey line/area _____
Name of chief surveyor _____   Date _____

Waypoint #	Date	Location Name	GPS Coordinate	PDOP	Ave. Diff. Y/N	Notes	Survey Person

APPENDIX B

CONVERTING SLOPE DISTANCE TO HORIZONTAL DISTANCE

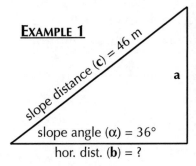

EXAMPLE 1

slope distance (**c**) = 46 m

a

slope angle (α) = 36°

hor. dist. (**b**) = ?

cosine of 36° = 0.8090
hor. dist. (b) = 46 × 0.8090
= 37.2 m

Slope (degrees)	Cosine	Slope (degrees)	Cosine	Slope (degrees)	Cosine
10	0.9848	22	0.9272	34	0.8290
11	0.9816	23	0.9205	35	0.8192
12	0.9781	24	0.9135	36	0.8090
13	0.9744	25	0.9063	37	0.7986
14	0.9703	26	0.8988	38	0.7880
15	0.9659	27	0.8910	39	0.7771
16	0.9613	28	0.8829	40	0.7660
17	0.9563	29	0.8746	41	0.7547
18	0.9511	30	0.8660	42	0.7431
19	0.9455	31	0.8572	43	0.7314
20	0.9397	32	0.8480	44	0.7193
21	0.9336	33	0.8387	45	0.7071

The slope distance that you measured with a metre tape in the field must be corrected, or flattened, to horizontal distance before you convert it to map distance and draw it on a flat map. Look up the cosine of the slope on the correct chart, depending on whether you measured slope in degrees or percent. Then use this equation:

Horizontal distance = slope distance × cosine of slope

where the slope distance is the distance measured in the field and the slope is the angle of the slope (also from the field).

Here are several additional equations related to slope:
α is the slope angle (degrees)
a is the height gain
b is the horizontal distance
c is the slope distance
S is the slope in percent
s is the slope in percent / 100
sqrt means 'square root'

$\cos α = 1 / \sqrt{s^2 + 1}$
$\cos α = b / c$
$S = 100 × \tan α$
$s = \tan α = a / b$
$b = c × \cos α$
$c = b / \cos α$
$c = \sqrt{a^2 - b^2}$
$b = \sqrt{c^2 - a^2}$
$\sin α = a / c$

EXAMPLE 2

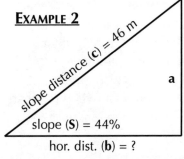

slope distance (**c**) = 46 m

a

slope (**S**) = 44%

hor. dist. (**b**) = ?

cosine for 44% = 0.9153
hor. dist. (b) = 46 × 0.9153
= 42.1 m

Slope (percent)	Cosine	Slope (percent)	Cosine	Slope (percent)	Cosine
10	0.9950	46	0.9085	76	0.7962
15	0.9889	48	0.9015	78	0.7885
20	0.9805	50	0.8944	80	0.7809
22	0.9766	52	0.8872	82	0.7733
24	0.9724	54	0.8799	84	0.7657
26	0.9678	56	0.8725	86	0.7582
28	0.9630	58	0.8650	88	0.7507
30	0.9578	60	0.8575	90	0.7433
32	0.9524	62	0.8499	92	0.7359
34	0.9468	64	0.8423	94	0.7286
36	0.9409	66	0.8346	96	0.7214
38	0.9348	68	0.8269	98	0.7142
40	0.9285	70	0.8192	100	0.7071
42	0.9220	72	0.8115		
44	0.9153	74	0.8038		

APPENDIX C

CONVERTING HORIZONTAL DISTANCE TO MAP DISTANCE

Below is a simple way to convert the distance that you measured in metres on the ground (and corrected for slope) to the distance that you want to draw on the map in centimetres.

map distance (cm) = horizontal distance (m) / divider

MAP SCALE	DIVIDER
1:500	5
1:1000	10
1:1500	15
1:2000	20
1:3000	30
1:5000	50
1:10,000	100
1:20,000	200
1:50,000	500
1:100,000	1000

EXAMPLE 1

horizontal distance = 37.2 m
map scale = 1:3000
map distance = 37.2 m / 30
= 1.24 cm

EXAMPLE 2

horizontal distance = 68 m
map scale = 1:2000
map distance = 68 m / 20
= 3.4 cm

SELECTING MAP SCALE FOR 1-METRE WIDE PAPER

WIDTH OF GROUND AREA (km)	SCALE
0.5–0.9	1:1000
1–1.9	1:2000
2–2.9	1:3000
3–3.9	1:4000
4–4.9	1:5000
5–9.9	1:10,000
10–19.9	1:20,000
20–24.9	1:25,000
25–49.9	1:50,000
50–100	1:100,000

First estimate the size of the area that you want to map, then check the table to see which scale is most suitable. (The amount of space for the margin will range from 5 cm to 25 cm.) If you are using precut sheets of paper—and not cutting them off a roll yourself—be sure to also check that the length of the ground area, when reduced to the same scale, will fit the length of the sheets available.

APPENDIX D

CALCULATING AREA FROM A MAP

GRAPH PAPER METHOD

Place translucent centimetre/millimetre graph paper over the area on the map (such as a field) and count the number of square centimetres within the boundary. Around the edges of the area you will see partial square centimetres, so count square millimetres there. Then use the multipliers below that correspond to the scale of your map.

MAP SCALE	HECTARES IN 1 cm^2	HECTARES IN 1 mm^2
1:1000	0.01	0.0001
1:2000	0.04	0.0004
1:3000	0.09	0.0009
1:4000	0.16	0.0016
1:5000	0.25	0.0025
1:10,000	1.00	0.0100
1:20,000	4.00	0.0400
1:25,000	6.25	0.0625
1:50,000	25.0	0.250
1:100,000	100.	1.00

EXAMPLE

Suppose that you want to measure the shaded area on the map to the right and the scale is 1:5000. First count up the square centimetres (indicated by the large '1's), then count up all the extra square millimetres around them. For ease of counting, count the squares in each of the little areas and then add up the subtotals (smaller figures).

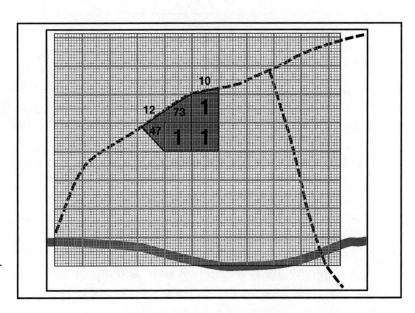

Number of $\text{cm}^2 = 1 + 1 + 1 = 3$

Number of $\text{mm}^2 = 73 + 47 + 12 + 10$
$= 142$

Total area = sum of hectares in square centimetres + sum of hectares in square millimetres
$= (3 \times 0.25) + (142 \times 0.0025) = 1.105 \text{ ha}$

APPENDIX E

CONVERTING LATITUDE AND LONGITUDE TO MAP DISTANCE

The map distance between successive minutes of latitude and longitude varies with latitude. *The first table is an example from Borneo. Do not try to apply these figures at significantly more northerly or southerly locations (more than 10° from the equator), because they are increasingly inaccurate the farther you get from the equator.*

MAP SCALE	CENTIMETRES IN 1 MINUTE	CENTIMETRES IN 10 SECONDS	MILLIMETRES IN 1 SECOND
1:10,000	18.4	3.1	3.1
1:20,000	9.2	1.5	1.5
1:50,000	3.7	0.6	0.6

The following table gives values for a 1:10,000 map at different latitudes so that you can see how the number of centimetres per minute varies. You could ask a cartography specialist or a cartography student if you need more precise information for your region.

LATITUDE (N OR S)	MERIDIAN (MEASURED N–S)	PARALLEL (MEASURED E–W)	LATITUDE (N OR S)	MERIDIAN (MEASURED N–S)	PARALLEL (MEASURED E–W)
0°	18.4	18.6	50°	18.5	12.0
10°	18.4	18.3	60°	18.6	9.3
20°	18.5	17.4	70°	18.6	6.4
30°	18.5	16.1	80°	18.6	3.2
40°	18.5	14.2	90°	18.6	0.0

Both halves headed: CENTIMETRES IN 1 MINUTE AT 1:10,000 SCALE

CONVERTING UTM MEASUREMENTS TO MAP DISTANCE

With UTM, the relationship between coordinates and map distance is consistent throughout the range of the UTM projection (80°S to 84°N; the UPS projection is used for polar regions). UTM coordinates are in metres, so distance conversion is usually just a matter of subtracting coordinates and accounting for the map scale. However, if your measurements cross the boundary from one 100,000 m square to another, be sure to take that into account (if your GPS receiver doesn't automatically take care of it for you).

Note: Although your GPS receiver will probably give you coordinates with all the digits the same size, in the margin of a UTM map with a 1000 m grid (for example, at 1:50,000) you might see UTM coordinates shown like this: 465000E, 5772000N. The big numerals give the kilometres and tens of kilometres. With a 10,000 m grid (for example, at 1:250,000), only the numeral giving the tens of kilometres is large (for example, 5770000N).

MAP SCALE	CENTIMETRES PER 1000 METRES
1:10,000	10.0
1:20,000	5.0
1:50,000	2.0

EXAMPLE

The east–west distance between 572500E and 568300E is 4200 m. On a map with a scale of 1:20,000, that corresponds to 21 cm.

APPENDIX F

STORING MAPS

Museums and archives store maps flat. For long-term storage, this way is the best. However, you'll need specially built map drawers—large, flat drawers that will take up space in your office. If you are using the maps a lot, then you'll inevitably have a struggle getting to the maps at the bottom of the stack and later re-filing them. Another way of storing maps flat is to hang them on a rack.

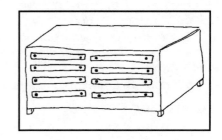

If you don't have the space to store maps flat, you can roll or fold your maps. Whether to fold or to roll is always a big question. Some materials, such as tracing paper and mylar, must be rolled, because they break when folded. Original maps that you want to reproduce should always be rolled because fold lines will mark and distort the copies. If you roll your maps, develop a system for storing and marking them so that you can easily find the one you are looking for. Write the title on an extra piece of paper and attach it to the outside of the roll. Write the title on a corner (or even a corner at each end, horizontally) of the back of the map so that you can see it when it is rolled. Store rolled maps in rigid tubes or in a supportive structure so that the ends too are protected. You can easily build a wooden map rack.

Although the original map is stored flat or rolled, paper maps that you are using a lot, such as field maps or reference maps, are better folded (usually to one or several standard sizes). Folded maps take less space to store and are easier to handle in some situations. Here are three rules for folding maps:

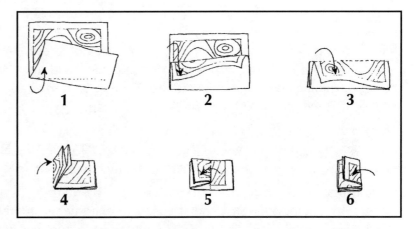

❖ *Fold with the printed side out, so that you don't have to unfold the whole map whenever you want to look at one part of it. (You can keep it in a plastic bag or an envelope to prevent soiling.)*

❖ *Fold with the title facing out at the top so that you can find the right map easily in a pile.*

❖ *To minimize wear at the fold lines, make all the folds opposite rather than doubled-up—that is, 'accordion' or zig-zag style rather than 'wrapper' style.*

APPENDIX G

SOURCES OF MAP DATA

Some sources of maps, photographs, and images include
- ❖ *government departments*
- ❖ *forestry companies or other companies*
- ❖ *development agencies and projects*
- ❖ *universities*
- ❖ *non-governmental organizations (NGOs)*

GOVERNMENT DEPARTMENTS

In most any country, a large number of government departments produce maps at the national, regional, and district levels. Some of these government bodies include

- ❖ **Lands, surveys, and mapping departments:** Usually produce detailed, large scale topographic maps that can be ordered. These agencies also have remote sensing images such as aerial photographs, radar images, or satellite images, but in many countries these products can be obtained only with security clearance.

- ❖ **Agencies for planning regional development:** Often have, or have access to, maps at the provincial and district level.

- ❖ **Military mapping units:** Produce topographic maps.

- ❖ **Resource ministries such as forestry and agriculture:** Have maps for the national, regional, and district level. However, they are not always willing to share this information.

In theory, anybody can apply for maps, photographs, or images through official government bodies. The procedure for applying is to write a letter to the government organization, describing what type of map is required and how the map will be used. However, before applying for certain maps (usually maps or imagery that have been taken from an airplane) it will often be necessary to apply for security clearance.

FORESTRY COMPANIES OR OTHER COMPANIES

If forestry companies are working in a specific area of forest, they will usually have good-quality, detailed maps of the area. Other companies working in the area, such as agribusinesses or mining companies, should also have maps. It might be worth trying to ask them at a local office if they will lend or give you a map.

DEVELOPMENT AGENCIES

If foreign development agencies (such as DFID, CIDA, USAID, EEC, or GTZ) are working in your area, they may also have maps or know how to obtain them.

UNIVERSITIES

Geography, geology, forestry, and agriculture departments at universities may also have maps or know how to obtain them, especially if the department has being doing any work in the area for which the map is wanted.

NON-GOVERNMENTAL ORGANIZATIONS

Small local non-governmental organizations (NGOs) and larger national and international NGOs (such as WWF-Indonesia) may also have maps or know how to obtain them.

—adapted from Manual on Participatory Resource Inventory
by Mary Stockdale and Jon Corbet

GLOSSARY

Note: underlined words are defined elsewhere in this glossary.

Absolute location: A point on the Earth's surface expressed by a coordinate system, such as latitude and longitude, or UTM.

Aggregation: A form of generalization that involves representing several similar nearby features (such as gardens or rock-piles) as a single, larger feature on a map.

Air (aerial) photographs: Remote-sensing photographs taken from an airplane.

Almanac: A continuously updated collection of data that a GPS receiver uses to determine the positions of the GPS satellites when it calculates coordinates.

Area feature: Something on the land—such as a plantation, hunting area, marsh, or lake—large enough to be depicted at scale on a map (shown with a polygon).

Attribute data: Information about a feature on a map, or thematic information.

Azimuth: The angle (often in degrees) that a certain direction (to a landmark, for example) is from the north meridian at a certain place.

Base map: A map that contains geographical reference information on which attribute data may be plotted to make thematic maps.

Bearing: A directional measurement taken by an observer, or the measured angle (often in degrees) between the north meridian and the line joining the observer and the object. Directions or azimuths are bearings.

Blueprinting: An inexpensive method for replicating black-and-white drawings, such as maps, that have been drawn on translucent paper, through the use of a *blueprinting machine*.

Cartography: The art or science of making maps.

Clinometer: A device for measuring slope angles.

Compass: A device for indicating direction, traditionally by the alignment of a magnetic needle that pivots to align with the direction of the Earth's magnetic field, though some recent models use electronic circuitry instead.

Compass Survey: See 'Traverse.'

Contour (line): On a map, a line that joins places of equal height above sea level. On a given map, contour lines are normally at specific increments, such as 25 m or 40 m, depending on the scale and the terrain.

Conversion: A form of generalization that involves changing the way a feature (or group) is represented—for example, several point features may be represented with a polygon, or a long, thin area may be depicted by a line symbol.

Coordinate: A pair of numbers that gives the location of a particular place on the Earth's surface in relation to a coordinate system, such as latitude and longitude or UTM.

Coordinate system: A pattern or network of crossing lines by which a position may be determined.

Database: A collection of information, for instance, about a particular community. A database is most useful if it is well organized and indexed.

Datum: A point from which other things are measured. This term (in full, *geodetic datum* or *geocentric datum*) can also refer to a cartographical system (specifically, a *reference ellipsoid*, such as WGS84) that is used to mathematically correct for irregularities in the Earth's sphericity, such as when using the GPS.

Declination (variation): The angle between the magnetic north meridian and the true north meridian at any given location. It is said to be 'east' or 'west' by a certain number of degrees according to whether the magnetic north meridian is east or west of the true north meridian.

Degree: A unit (abbreviated as °) for measuring direction as if from the centre of a circle. There are 360 degrees in a circle. Each degree can be subdivided into 60 minutes (abbreviated as '). Each minute can be divided into 60 seconds (abbreviated as "). Bearings and declination, for example, are usually (but not always) measured in degrees.

Differential GPS: A method of correcting for errors in GPS coordinates by using two receivers, one to rove and collect position data, the other to remain stationary at a known position to collect correction data that is transmitted to the roving receiver (or supplied to it at a later time).

Digitize: To convert an image, such as a map, into a form that a computer can store and manipulate through the use of special software (a computer program). Digitizing is usually done manually, with a *digitizing tablet*, but simply *scanning* the image may be suitable for some purposes.

Displacement: A form of generalization that involves moving close-set map symbols slightly out of their correct locations so that they do not overlap each other.

Dissolution: A form of generalization that involves combining two or more adjoining polygons representing somewhat different kinds of features into one polygon.

Easting: The part of a coordinate (such as longitude) that gives the east–west position.

Enhancement: A form of generalization—the opposite of simplification.

Ephemeral map: A temporary map, such as a ground map, intended to be kept for a short time only before it is destroyed.

Ephemeris: A map and calendar of the movement of celestial bodies or satellites.

Equator: The great circle (0° latitude) that connects all points that are at an equal distance from the north and south poles.

Feature: A definable and relatively permanent thing on the land (such as a house, boulder, hill, river, road, boundary, field, forest type, hunting area, sacred site, etc.) that can be depicted on a map.

Field: To go into or to be in 'the field' refers to doing a field survey or field-checking a map.

Field-check: To verify the locations of features shown on a map by going out onto the land and observing—and possibly measuring—their relationships to other features.

Field survey: To go out on the land to observe the features and draw a map based on first-hand observation—rather than drawing it from memory or descriptions or by interpreting remote-sensing data (see table-top mapping).

Frame: A rectangle in which a map or a map part, such as a legend, will be (or has been) drawn. Also, a drawing of a traverse that shows just the stations and the lines that join them, without sideshots or other details.

Generalization: The choosing of <u>features</u> and the method of their depiction in order to draw a clear and meaningful <u>map</u>; aspects of generalization include <u>aggregation</u>, <u>conversion</u>, <u>displacement</u>, <u>dissolution</u>, <u>enhancement</u>, <u>selection</u>, <u>simplification</u>, and <u>smoothing</u>. The degree and kind of generalization should be consistent throughout any given map.

Geographic coordinate system: The <u>grid</u> system of <u>latitude</u> and <u>longitude</u>.

Georeferenced: Refers to a map or photo that has been geographically corrected, so that every point on it shows <u>absolute location</u>. For example, air photos and <u>satellite</u> images are georeferenced to correct for scale distortions inherent in the process of collecting data through <u>remote sensing</u>.

GIS (Geographic Information System): A computerized system for the collection, storage, and retrieval of geographic data.

GPS (Global Positioning System): A system of artificial <u>satellite</u>s and ground units that enables a user with a portable receiver to determine <u>absolute location</u>s with good accuracy.

Gradian (also called 'gon' or 'grade'): A unit of angular measure, an alternative to <u>degree</u>s. There are 400 gradians in a circle (100 in a right angle), so one gradian = 0.9°.

Graph paper: Paper printed with a pattern of intersecting lines parallel to the edges and at fixed increments (such as 5 mm or 1 mm).

Graph scale: A graphic representation of <u>map</u> <u>scale</u> proportions using a bar and numbers to indicate distance.

Grid: A pattern or network of crossing lines (such as on a <u>map</u>) by which a position may be determined.

Grid north: North as indicated by the <u>north meridian</u>s of a particular <u>map projection</u>.

Ground map: A large and temporary <u>map</u> (perhaps 10 m × 10 m in size), constructed outside on the ground using leaves, rocks, beans, wood, reeds, or other materials, or created indoors using hats, shoes, rope, pieces of paper, etc.

Hip chain: A measuring tool, used in <u>survey</u>ing and worn on a belt, that consists of a small plastic box containing a roll of thread. Pulling thread out of the box operates a counter that reads distance in metres and tenths of a metre.

Horizontal distance: Distance along the horizontal (as distinguished from <u>slope distance</u>).

Index: An alphabetical list of <u>keyword</u>s that indicates where in a book or <u>database</u> each topic is discussed or mentioned.

Index contour: A <u>contour line</u> that is darker or thicker than the regular ones to assist in more quickly determining elevation. Index contours usually fall every fifth (or fourth) line and represent round-number elevations, such as 250 or 500 m.

Information unit: A piece of information; for example, a story transcript, photograph, video-tape, etc. that contains or depicts knowledge about the community, its land, its people, and its history.

Intersection: A <u>survey</u> technique that involves taking <u>bearing</u>s from two known places to identify the location of a third, unknown location.

Keyword: A significant word (subject name or topic) that is used in <u>index</u>ing a collection of information (<u>database</u>) to make it easier to find specific pieces of information.

Landmark: An obvious feature in the landscape.

LANDSAT: A specific kind of <u>satellite</u> image that shows a larger area than a <u>SPOT</u> image.

Latitude: Parallel lines running east-west around the globe; measured in degrees north or south from the equator.

Legend: The part of a map (or an additional sheet) that explains what the symbols on the map mean.

Light table: A piece of drafting equipment that consists of a translucent work surface (with or without legs) with a light source beneath it, used to facilitate the copying of information from one sheet of paper (or plastic) to another.

Line feature: Something on the land that is relatively long and thin—such as a river, road, trail, or boundary; its symbol on a map may be exaggerated in width if it would otherwise be too narrow to show at scale.

Location map: A small, small-scale map that shows where the land depicted on the main map is in relation to the whole state, province, or country.

Longitude: Meridian lines running north–south and joining at the poles; measured in degrees from the Prime Meridian (0°).

Magnetic dip: The angle at which the Earth's magnetic field at a particular place would tilt a freely suspended magnetic needle relative to the horizontal. Some types of compass can and should be mechanically adjusted for use in different regions of the world.

Magnetic north: The direction of the meridian along which a freely suspended magnetic needle would lie if it were influenced only by the Earth's magnetic field. Magnetic north is constantly moving, albeit so slowly that in almost all locations this movement causes only negligible error in compass use.

Map: A picture of the land, a map is a graphic representation, often two-dimensional, of some part (or all) of the Earth's surface. There are many different kinds of maps.

Map interview: The process of talking to community members and asking questions to help them record their information about the land on a map, in sketch form, or in words.

Map projection: A particular way (such as UTM) of depicting the curved surface of the Earth as a two-dimensional map through the use of a specific mathematical algorithm.

Map series: A set of thematic maps of the same area, or a set of maps (that were made with the same process and format) to cover a region too large to fit on one map sheet at the desired scale.

Map registration: A technique by which to align two or more maps, such as an overlay map and a base map, using special registration marks (or special holes and pins).

Mental map: A map that represents the perceptions and knowledge that a person has of an area.

Meridian: A *great circle* around the Earth, or half of one. A *meridian of longitude* (or line of longitude) connects the north and south poles. The meridian of longitude that passes through any particular point can be called the *north meridian* for that point.

Metre tape: A measuring tool used in surveying that is marked in metres. Basic models consist of a rolled nylon or fibreglass tape that extends to 30 or 50 m.

Mylar: A particular kind of drawing 'paper' made of plastic. It is available with one or both sides 'frosted' (matte) to take pencil or drawing ink.

NGO (Non-Governmental Organization): An organization, usually with humanitarian or environmental protection objectives, that is not controlled by a government, though it may operate with the assistance of government funding. Many NGO projects are intended to aid indigenous peoples to protect or improve their quality of life.

North line: A line drawn on a map so as to align with a north meridian. It provides a reference line by which to measure bearings by using a compass or protractor.

Northing: The part of a coordinate (such as the latitude) that gives the north–south position.

Offset: The perpendicular distance from a traverse line to a parallel line or to a point.

Orientation: The positioning of a map so that its north line points to the Earth's true north.

Overlay map: A thematic map on tracing-paper (or on a plastic sheet) that is used in conjunction with a base map.

Panorama sketch: A landscape sketch made from a location that has a view of the surrounding terrain for a fair distance.

Parallel (of latitude): A circle on the Earth's surface that is parallel to the equator, but smaller and either to the north or south of it. A line of latitude.

PDOP (Precision Dilution of Position): PDOP is an estimate of the accuracy of a GPS position fix based on the quality of the satellite signals (which is a result of the satellite distribution at the time of the determination).

POC (Point of Commencement): The starting point for a survey route.

Point feature: Something—such as a sacred rock, house, or special tree—that is too small to be drawn to scale on a particular map, so it is instead represented by a standardized symbol that may be either abstract or stylized.

Polygon: A bounded area on a map that represents something (an area feature such as a lake, field, forest type, or hunting area) on the land that is large enough to be shown to scale. A polygon can be identified through the use of a particular colour, pattern, or code.

Position Averaging: A method for improving the accuracy of GPS data that requires just one GPS receiver, which is set up to take a series of readings over a period of time.

Prime meridian: Zero degrees longitude. Also known as the Greenwich Meridian because it was established at the Greenwich Observatory near London, England.

Projection: See 'Map projection.'

Protractor: A device, usually of clear plastic and circular or D-shaped, used to measure angles.

PRA (Participatory Rural Appraisal): A set of techniques for including the local people in the documentation and analysis of local land issues.

Radian: A unit of angular measure, an alternative to degrees. There are $2 \times$ pi radians in a circle, so one radian is approximately $57.3°$.

Reference map: A base map that has been made more locally relevant by ground-checking (and correcting if necessary) major features and adding local landmarks and place names. 'Reference map' (or *reference base map*) may refer specifically to the final base map on which all the information from field surveys and other sources has been compiled.

Registration marks: Small marks (usually '+' symbols) used to simplify the aligning of two or more maps (such as tracing paper or plastic thematic maps on top of a base map) so that the features on the top map(s) are in their correct positions with respect to the features on the bottom map.

Relative location: A location of a place in relation to (for example, 600 m southwest of, or 100 m downhill from) another place (usually one whose absolute location is already known).

Remote sensing: The process of gathering information about the Earth from a distance. Such data is commonly gathered by satellite or air (aerial) photography.

Resection: A survey technique that involves taking underline{bearing}s to two known places to determine the location of a third, unknown location at which you are standing.

Resolution: The smallest distance or size of object that can be seen in an image (as acquired, for instance, through remote sensing).

Satellite: A platform launched into close orbit around the Earth and used to carry electronic equipment that transmits information back to Earth. Some satellites are used to transmit pictures of the Earth from space back to Earth for remote sensing applications. The GPS uses 24 satellites that were made and launched specifically for transmitting signals to GPS receivers on Earth.

Selection: A form of generalization that involves choosing which of a number of underline{feature}s (or which parts thereof) to show on a map, while omitting others.

Selective availability: A procedure by which the United States Department of Defense (USDoD) deliberately and intermittently interferes with the signals from GPS underline{satellite}s so that civilian and other non-USDoD GPS receivers cannot calculate extremely precise locations, but their own units can. The errors thus introduced must be taken into account (and perhaps strategically minimized) by civilian GPS users.

Scale: The relationship between distance on a map and on the Earth's surface, usually represented as a ratio (for example, 1:10,000) or with a graph scale.

Sideshot: Along a survey route, a short branch or spur made for the purpose of accurately recording an important feature located a short distance to one side of the route.

Simplification: A form of generalization that involves deleting some of the underline{survey}ed points that show the path of a line feature or the boundary of a polygon so as to remove excessive detail.

Sketch map: A drawing of a place or area, not drawn with accurate or measured scale or direction. A *scale sketch map* is a sketch given scale by fitting it onto a topographic map, without a field survey.

Smoothing: A form of generalization that involves averaging (either by visual estimation or computation) the locations of the underline{coordinate}s that define the underline{survey}ed path of a line feature or the boundary of a polygon so as to remove excessive detail, given the scale of the map, or to average measurement errors.

Slope Distance: A distance measured on sloping terrain that has not yet been converted to horizontal distance for plotting on a survey drawing or map.

SPOT: A specific kind of satellite image that covers a smaller area than a LANDSAT image and at a higher resolution (and usually at a higher cost per square kilometre).

Spot height: The exact height, shown with a number on a map, of a particular place above some datum (usually mean sea level).

Station: A starting point or endpoint of a survey leg. Stations are where measurements of distance and bearings are taken and recorded, along with any relevant notes. The stations within each surveying project are sequentially numbered for identification.

Stereoscope: A device used to look at paired air (aerial) photographs, making it possible to see features on the photographs in three-dimensional perspective.

Survey: To traverse a particular linear feature (such as a boundary or a river), or travel in some specific pattern across a particular area, with the purpose of recording the locations of features on the land and details about them for use in making a map. Surveying is often done with a compass and a metre tape; some surveys are done with a GPS receiver.

Survey chain: A surveying tool that consists of a nylon rope on which every tenth of a metre is marked by a metal clip.

Table-top mapping: The drawing of a map—or the addition of thematic information to an existing base map—using information from memory or from remote sensing or photographs or notes, rather than while actually out on the land doing a field survey.

Thematic map: A map that depicts specific themes or sets of information; for example, forest type, land use, historical migration, property ownership, or animal habitat.

Three-dimensional (3-D): Refers to a map, such as a cardboard relief map that extends above its base according to the height of the land—or to the image seen through a stereoscope.

Topographic map: A contour map that shows human-made and natural physical features. A topographic map at a scale of 1:10,000 to 1:50,000 would be a good base map.

Topography: The shape or configuration of the Earth's surface; used especially in regards to the part of it within visual range from some particular place.

Tracing paper: A lightweight and translucent drawing paper that allows the copying of images that can be seen through it.

Transect: Surveying in a straight line across the land, usually for the purpose of mapping or recording information along the line. Transects are often conducted for a resource inventory.

Transect sketch: A sketch map made by observing and drawing the features seen on both sides of the route as the mapper performs a transect. It can be from a bird's-eye perspective or a profile perspective.

Traverse: A survey done by walking along the ground with a compass and metre tape. The four types used in community mapping are *linear*, *boundary* (*closed*), *grid*, and *radial*.

Triangulation: A survey technique to find the location of an 'unknown' position on a map by using bearings to (or from) three known locations.

Type line: The outline (boundary) of a polygon drawn on a map.

UPS (Universal Polar Stereographic): A common map projection and grid system for the polar regions (poleward of 80°S and of 84°N) that is used in conjunction with the Universal Transverse Mercator grid.

UTM (Universal Transverse Mercator): A common map projection and grid system for the part of the Earth's surface between 84°N and 80°S that is widely used for topographic maps, air (aerial) photographs, and satellite images. It divides this area into 1200 zones (each identified by a unique number-letter code, such as 28M) that are further subdivided and coded.

Variation: See 'Declination.'

Watershed: The area that a certain river or lake and all its tributaries drain.

Waypoint: A surveying term used to describe a 'position fix' (the coordinates) of a place, especially if determined through the use of a GPS receiver. The waypoints in any given surveying project are sequentially numbered.

REFERENCES USED AND BOOKS ABOUT MAPPING

Aberley, Doug. 1993. *Boundaries of Home: Mapping for Local Empowerment.* New Society Publishers, Gabriola Island, BC, Canada.

Alcorn, Janis B. 2000. *Borders, Rules and Goverance: Mapping to Catalyze Changes in Policy and Management.* Gatekeeper Series, no. 91. International Institute for Environment and Developement, London, UK.

Alcorn, Janis B. 2000. 'Keys to Unleashing Mapping's Good Magic.' *PLA Notes*, no. 39, 10–13. Institute for Environment and Developement, London, UK.

Brody, Hugh. 1981. *Maps and Dreams.* Douglas and McIntyre, Toronto, Canada.

Burroughs, P.A. 1986. *Principles of Geographic Information Systems for Land Resources Assessment.* Oxford Science Publications, Oxford, UK.

Carson, Brian. 1985. 'Appraisal of Rural Resources Using Aerial Photography: An Example from a Remote Hill Region in Nepal.' Paper presented at the International Conference on Rapid Rural Appraisal, September 2–5, 1985, in Khon Kaen, Thailand.

Crane, Julia and Michael Angrosino. 1984. *Field Projects in Anthropology: A Student Handbook.* Waveland Press, Prospect Heights, IL, USA.

Cultural Survival Quarterly magazine, winter 1995. 'Geomatics: Who Needs It?' Cambridge, MA, USA.

Denniston, Derek. 1994. 'Defending the Land with Maps.' *Worldwatch*, 7(1):27–31.

Dreyfuss, Henry. 1972. *An Authoritative Guide to International Graphic Symbols.* McGraw-Hill, New York, USA.

Eastman, Ronald J. 1992. *IDRISI User's Guide.* Clark University, Graduate School of Geography, Worcester, Massachusetts, USA.

Fox, Jefferson. 1990. 'Diagnostic Tools for Social Forestry.' In *Keepers of the Forest: Land Management Alternatives in Southeast Asia*, edited by M. Poffenberger. Kumarian Press, West Hartford, CT, USA.

Freudenberger, Karen Schoonmaker. 1995. *Tree and Land Tenure: Using Rapid Appraisal to Study Natural Resource Management.* FAO, Rome, Italy.

Gallagher-Mackay, Kelly. 1996. 'Maps, Knowledge and Territory: International.' *Review of Current Law and Law Reform* 2:8–17.

Greenwood, David. 1964. *Mapping.* University of Chicago Press, Chicago, USA.

Grenier, Louise. 1998. *Working with Indigenous Knowledge: A Guide for Researchers.* International Development Research Centre, Ottawa, Canada.

Gupta, A.K. and IDS Workshop. 1989. 'Maps Drawn by Farmers and Extensionists.' In *Farmer First: Farmer Innovation and Agricultural Research*, edited by R. Chambers, A. Pacey, and L.A. Thrupp. Intermediate Technology Publications, London, UK.

Harrington, Sheila, ed. 1999. *Giving the Land a Voice: Mapping Our Home Places.* Land Trust Alliance of British Columbia, Saltspring Island, Canada.

Hulbert, John. 1973. *All Asbout Navigating and Route Finding.* Carousel Books, London, UK.

Hurn, Jeff. 1989. *GPS: A Guide to the Next Utility.* Trimble Navigation, Sunnyvale, CA, USA.

Johnson, Martha, ed. 1992. *Lore: Capturing Traditional Environmental Knowledge.* International Development Research Centre, Ottawa, Canada.

Jones, P. Alun. 1968. *Field Work in Geography.* Longmans, Harlow, UK.

Khan, Asmeen, Chun K. Lai, Ruby Qudir, and S. Iqbal Ali. 1990. *Handbook on Land-use Mapping Resources and Sketch-mapping Techniques for Bangladesh.* Based on a workshop organized by Winrock-ADAB-IIESDM.

Mason, Adrienne. 1998. 'Mapping: Communities Discover That the Language of Cartography Speaks Louder Than Words.' *Canadian Geographic,* Sept./Oct., 58–62.

Milton Freeman Research Ltd. 1976. *Inuit Land Use and Occupancy Project. Vol. III: Land Use Atlas.* Department of Indian and Northern Affairs, Ottawa, Canada.

Monmonier, Mark. 1996. *How to Lie with Maps.* The University of Chicago Press, Chicago, USA.

Peluso, Nancy Lee. 1994. 'Whose Woods Are These? The Politics of Mapping State and Indigenous Forest Territories in Kalimantan, Indonesia.' Presented to the Annual Meeting of the American Association of American Geographers, San Francisco, USA.

Poole, Peter. 1995. *Land-based Communities, Geomatics and Biodiversity Conservation: A Survey of Current Activities.* Biodiversity Support Programme, World Wildlife Federation, Washington, USA.

Poole, Peter. 1995. *Indigenous Peoples, Mapping & Biodiversity Conservation: An Analysis of Current Activities and Opportunities for Applying Geomatics Technologies.* Biodiversity Support Program, World Wildlife Federation, Washington, USA.

Pretty, Jules, Irene Guijt, Ian Scoones, and John Thompson. 1995. *A Trainer's Guide for Participatory Learning and Action.* Sustainable Agriculture Program, International Institute for Environment and Development, London, UK.

Rundstrom, Robert A. 1993. 'The Role of Ethics, Mapping, and the Meaning of Place in Relations Between Indians and Whites in the United States.' *Cartographica* 30 (Spring): 21–28.

Stockdale, M.C., and J.M.S. Corbett. 1999. *Participatory Inventory: A Field Manual Written with Special Reference to Indonesia.* Tropical Forestry Papers, no. 38, Oxford Forestry Institute, University of Oxford, UK.

Toledo Maya Cultural Council. 1997. *Maya Atlas: The Struggle to Preserve Maya Land in Southern Belize.* North Atlantic Books, Berkeley, USA.

Turnbull, David. 1994. *Maps are Territories: Science in an Atlas: A Portfolio of Exhibits.* University of Chicago Press, Chicago, USA.

Tyner, Judith. 1992. *Introduction to Thematic Cartography.* Prentice-Hall, New Jersey, USA.

Union of BC Indian Chiefs. 'Alaska Highway Pipeline Impact Study: Land-use and Occupancy Mapping Instructions.' BC, Canada.

Weinstein, Martin S. 1993. 'Aboriginal Land-use and Occupancy Studies in Canada.' A paper prepared for the Workshop on Spatial Aspects of Social Forestry Systems. East-West Center, Honolulu, USA.

Weinstein, Martin S. 1997. 'Getting to Use in Traditional Use Studies.' A paper presented to the Society for Applied Anthropology, Seattle, USA.

SOME INTERNET RESOURCES

Note: Websites change from time to time. If the page that you are looking for is no longer available at the address given, you can sometimes find it by using a search engine (*such as* **http://www.google.com** *or* **http://www.dogpile.com**) *to look for the article title or organization name.*

Aboriginal Mapping Network. Case studies, methodologies, articles, events, resources. **http://www.nativemaps.org**

Cartography Today magazine. **http://www.mapfacts.com**

Community Technology Centre Review. Article 'Mapping Community Resources.' **http://www.ctcnet.org/r981cor2.htm**

GPS World. Article 'Nicaragua's "GPSistas": Mapping their Lands on the Caribbean Coast.' **http://www.gpsworld.com/0199/0199feat.html**

Green Map System. Locally made maps, by mapmakers of all ages and in a variety of styles, that show urban ecology, with a goal of improving community sustainability. **www.greenmap.com**

International Institute for Rural Reconstruction. Online publication of the manual *Recording and Using Indigenous Knowledge.* **http://www.panasia.org.sg/iirr/ikmanual/**

New York Public Research Interest Group. Community mapping resources, mostly urban. **www.cmap.nypirg.org**

Participatory Avenues. Information on Integrated Approaches to Participitory Development articles as well as a forum for discussion on community mapping. **www.iapad.org**

Trimble Navigation. Basic tutorial on how the GPS works. **www.trimble.com**

University of Wisconsin, Geography Department. 'Resources for Cartography, GIS and Remote Sensing.' **www.uwsp.edu/acaddept/geog/cart.htm**